AF842842

Near-Field Microseismic Data Processing in Deep Tunnels

Yaxun Xiao · Shujie Chen · Liu Liu

Near-Field Microseismic Data Processing in Deep Tunnels

Methods and Applications

同濟大學出版社
TONGJI UNIVERSITY PRESS

Springer

Yaxun Xiao
Institute of Rock and Soil Mechanics
Chinese Academy of Sciences
Wuhan, China

Liu Liu
Institute of Rock and Soil Mechanics
Chinese Academy of Sciences
Wuhan, China

Shujie Chen
Department of Geological and Hydraulic Engineering
Tongji University
Shanghai, China

ISBN 978-981-92-3207-9 ISBN 978-981-92-3208-6 (eBook)
https://doi.org/10.1007/978-981-92-3208-6

Jointly published with Tongji University Press Co., Ltd.
The print edition is not for sale in China (Mainland). Customers from China (Mainland) please order the print book from: Tongji University Press Co., Ltd.

This work was supported by National Natural Science Foundation of China (12441510, 42477195).

© Tongji University Press Co., Ltd 2026

This work is subject to copyright. All rights are solely and exclusively licensed by the Publisher, whether the whole or part of the material is concerned, specifically the rights of translation, reprinting, reuse of illustrations, recitation, broadcasting, reproduction on microfilms or in any other physical way, and transmission or information storage and retrieval, electronic adaptation, computer software, or by similar or dissimilar methodology now known or hereafter developed.
The use of general descriptive names, registered names, trademarks, service marks, etc. in this publication does not imply, even in the absence of a specific statement, that such names are exempt from the relevant protective laws and regulations and therefore free for general use.
The publishers, the authors, and the editors are safe to assume that the advice and information in this book are believed to be true and accurate at the date of publication. Neither the publishers nor the authors or the editors give a warranty, express or implied, with respect to the material contained herein or for any errors or omissions that may have been made. The publishers remain neutral with regard to jurisdictional claims in published maps and institutional affiliations.

This Springer imprint is published by the registered company Springer Nature Singapore Pte Ltd.
The registered company address is: 152 Beach Road, #21-01/04 Gateway East, Singapore 189721, Singapore

If disposing of this product, please recycle the paper.

Competing Interests The authors have no competing interests to declare that are relevant to the content of this manuscript.

Contents

1 Introduction 1
- 1.1 Prologue 1
- 1.2 Book Outline 2
- References 4

2 Automatic Rock Fracture Signal Identification Method Based on Tunnel MS Data 7
- 2.1 Prologue 7
- 2.2 Methodology 8
  - 2.2.1 Manual Identification Process Review 8
  - 2.2.2 Digitalization Method 8
  - 2.2.3 Threshold Determination 11
- 2.3 Field Case 11
  - 2.3.1 Field Case Background 11
  - 2.3.2 Threshold in the Jinping Tunnel 12
  - 2.3.3 Case Analysis 12
  - 2.3.4 Performance Assessment 14
- 2.4 Discussion 15
  - 2.4.1 Factors That Determine Thresholds 15
  - 2.4.2 Further Application in a Drill and Blast Tunnel 16
  - 2.4.3 Strengths and Limits 16
- 2.5 Conclusion 17
- References 17

3 P-Wave Arrival Time Picking Method Based on Low Signal-to-Noise Ratio (SNR) Data Without Noise Filtering 19
- 3.1 Prologue 19
- 3.2 MS Monitoring in Tunnels 20
  - 3.2.1 Data from the Jinping Tunnel 20
  - 3.2.2 Source Wave Traveling Path Difference Versus Wavelength of the P-Wave 22
  - 3.2.3 Characteristics of Noise Signals 22

3.2.4 SNR of Rock-Fracturing Signals 24
3.3 Methodology 26
3.3.1 Instantaneous Phase Difference 26
3.3.2 The Process of IPDI_P 28
3.4 Performance Analysis 30
3.4.1 Noise Simulation 30
3.4.2 Simulation Model and Parameters 31
3.4.3 Typical Simulation Case 33
3.4.4 Identifying P-Wave Arrival Times in More Complicated MS Signals 36
3.5 Field Application 39
3.6 Discussion 41
3.7 Conclusions 42
References 43

4 Determining S-Wave Arrival Times for Near-Field MS Data in Tunnel 45
4.1 Prologue 45
4.2 The Process of IPDI_S 46
4.3 Numerical Validation 48
4.3.1 Numerical Simulation Method 48
4.3.2 S-Wave Arrival Time Picking Based on the IPDI Method 49
4.4 Field Study 51
4.4.1 Background 51
4.4.2 IPDI Analysis of an Example Rockburst Signal 52
4.4.3 Assessing the Accuracy of the IPDI Method 56
4.5 Discussion 60
4.6 Conclusion 63
References 64

5 Absolute Location Method for MS Event Clusters in Deep Tunnels Based on Coda Wave Interference 67
5.1 Prologue 67
5.2 The Characteristics of the Coda Wave in the Tunnel 69
5.2.1 Background 69
5.2.2 Determination of Coda Waves 70
5.2.3 The Estimated Length of the Coda Wave 71
5.2.4 Correlation Between MS Signals and Other Signals 73
5.3 Methodology 74
5.3.1 Coda Wave Interference 74
5.3.2 Using CWI for MS Event Cluster Location in Tunnels 76
5.4 Numerical Validations 78
5.4.1 Preparation for Numerical Testing 78

5.4.2 Numerical Test Results of the CWI Method 79
5.5 Field Study 81
5.5.1 The Velocity Determination Applicable to the CWI Method 81
5.5.2 An Example of CWI Analysis 82
5.5.3 Assessing the Accuracy of the CWI Method 83
5.6 Discussion 86
5.7 Conclusion 88
References 89

6 MS Source Location in Deep-Buried Tunnels Using the Updated Inverted Heterogeneous Velocity 93
6.1 Prologue 93
6.2 Methodology 95
6.2.1 Motivation 95
6.2.2 Overall Workflow 96
6.2.3 Surface Wave Inversion 99
6.2.4 AST in Deep Tunnels 100
6.2.5 Locating MS Sources with Heterogeneous Velocity Models 103
6.3 Numerical Study 104
6.3.1 Numerical Modeling of the Blasting Source 104
6.3.2 Determining t_0 of the Blasting Source 107
6.3.3 P-Wave Traveling Time Tomography While Drilling 109
6.3.4 MS Source Location 111
6.4 Case Study 113
6.4.1 Background 113
6.4.2 Inversion of the Velocity Model 114
6.4.3 Validation by the Blasting Source Location 121
6.5 Discussion 122
6.6 Conclusion 124
References 125

7 Analyzing the Stability of Rock Surrounding Deep Cross-Tunnels Using a Dynamic Velocity Field 127
7.1 Prologue 127
7.2 Methodology 129
7.2.1 Overall Workflow 129
7.2.2 Adjoint-State Traveltime Tomography 130
7.2.3 Stability Analysis with Dynamically Updated Mechanical Properties 131
7.3 Case Study 132
7.3.1 Tomography Results 133

7.3.2 Stability Analysis Results ... 137
7.4 Discussion ... 144
7.5 Conclusions ... 146
References ... 146

8 Stratified Double-Difference Dynamic Tomography of a 3D Velocity Field for a Deep Buried Large Underground Cavern ... 149
8.1 Prologue ... 149
8.2 Methodology ... 151
8.2.1 Stratified Double-Difference Tomography Method (SDD) ... 151
8.2.2 Double-Difference Traveltime Tomography ... 155
8.3 MS Monitoring Setup for the Jinping Underground Laboratory ... 158
8.4 Resolution Test ... 159
8.4.1 Setup for the Test ... 160
8.4.2 The Evaluation Criterion for Inversion ... 161
8.4.3 Velocity Model Inversion ... 162
8.4.4 Inversion Results and Analysis ... 162
8.5 Realistic Synthetic Dynamic Inversion Tests ... 166
8.5.1 Design of the Excavation Process ... 166
8.5.2 Set-Up for the Dynamic Test ... 167
8.5.3 Velocity Model Inversion with Biased Data ... 168
8.5.4 Results and Analysis ... 170
8.6 Discussion ... 172
8.7 Conclusion ... 176
References ... 177

9 Conclusion ... 181
9.1 Automated Identification of Rock Fracture Signals (Chap. 2) ... 181
9.2 P-Wave Arrival Time Picking Under Low Signal-to-Noise Ratios (Chap. 3) ... 182
9.3 S-Wave Arrival Time Identification (Chap. 4) ... 183
9.4 MS Event Cluster Location Using Coda Wave Interferometry (Chap. 5) ... 183
9.5 MS Source Location with Updated Heterogeneous Velocity Models (Chap. 6) ... 184
9.6 Stability Analysis of Deep Cross-Tunnels (Chap. 7) ... 185
9.7 Dynamic Tomography of Three-Dimensional Velocity Fields (Chap. 8) ... 185
9.8 Overall Contributions ... 186

10 The Core Code of Methods Proposed in This Book 187
10.1 Declaration 187
10.2 Automatic Rock Fracture Signal Identification in Tunnel 187
10.3 MS Event Cluster Location in Tunnels Based on Coda Wave Interference 191
10.4 Adjoint-State Travel Time Tomography 200
Reference 214

Initialism

2D	Two-dimensional
ACC	Accuracy rate
AIC	Akaike information criterion
AST	Adjoint-state traveltime tomography
CC	Cross-correlation
CWI	Coda wave interferometry
DD	Double-difference
EDZ	Excavation damaged zone
EEMD	Ensemble empirical mode decomposition
eps	Equivalent plastic shear strain
FFT	Fast Fourier Transform
FMM	Fast marching method
FWI	Full waveform inversion
HT	Hilbert transformations
IPDI_P	Instantaneous phase difference intensity of P-waves
IPDI_S	Instantaneous phase difference intensity of S-waves
LERR	Local energy release rate
MIS	Miss rate
MS	Microseismic
NMO	Normal moveout
PML	Perfectly matched layer
PRF	Pure rock-fracturing
PSO	Particle swarm optimization
RMS	Root mean square
SDD	Stratified double-difference
SDDSR	The shortest distance between different sensor rows
SNR	Signal-to-noise ratios
STA/LTA	Short-term average to long-term average

SWTPD	The source wave traveling path difference
TBM	Tunnel boring machine
TSP	Tunnel seismic prediction

Symbols

Because there are many symbols and variables involved in this book, the symbols in each chapter are independent.

Chapter 2

A_i	A trace signal, i is the sampling point ($i = 1, 2, \ldots, I$)
A_i^n	Normalized signal of A_i
C	Cluster center calculated by K-means
C^*	Threshold to filter signals
B_i	The modified waveform
y_j^1	The points that satisfy Eq. 2.2, where $j = 1, 2, \ldots, J$, constituting the set $Y_1 = y_j^1$
x_j^1	The positions of elements in Y_1 that has a corresponding equal value in B_i, constituting the set $X_1 = x_j^1$
y_k^2	The points satisfy Eq. 2.4, where $k = 1, 2, \ldots, K$, constituting the set $Y_2 = y_k^2$
x_k^2	The positions of y_k^2 in B are labeled x_k^2, constituting the set $X_2 = x_k^2$
L	The number of pulses in a signal
p_k	The positions of x_k^2 can be identified and labeled p_k
$y_{p_k+1}^{2*}$	The amplitude values inserted at positions $x_{p_k+1}^1$
Y_{2*}	Signal after inserting, consisting of the inserted points and y_k^2
P_x	The corresponding sampling point is extracted as P_x when Y_{2*}equals 1
S_x	The first points with an amplitude smaller than 0.2, by backward search from P_x
F_x	The first points with an amplitude smaller than 0.2, by forward search from P_x
Q	The ratio between F_x, P_x, and S_x
Q^*	The threshold of Q
ACC	The accuracy rate
MIS	The miss rate
N_m	The number of rock fracture signals identified manually

N_w	The number of rock fracture signals identified by the proposed method
$N_{m\&w}$	The number of rock fracture signals identified both manual and digital identification
N_o	The number of rock fracture signals omitted by the digitalization method
α	A suitability rate used to evaluate which threshold pair is optimal

Chapter 3

F_a	An average frequency
M	The number of MS signals
f_i	The main frequency in the i-th MS signal
L_P	The wavelength of P-wave
V_p	The propagation velocity of P-waves
$SNR_{Average}$	The average SNR of each event
$SNR_{standard}$	The standard deviation SNR of each event
SNR_i	The SNR of the i-th signal
l_A	The distant between MS source and sensor A
l_B	The distant between MS source and sensor B
t_{Aa}	The arrival times of signals recorded by sensor A
t_{Ba}	The arrival times of signals recorded by sensor B
t_{Ae}	The end times of the first periods of the P-waves at A
t_{Be}	The end times of the first periods of the P-waves at B
$\omega_A(t)$	The instantaneous phases of signal recorded by sensor A
$\omega_B(t)$	The instantaneous phases of signal recorded by sensor B
$\Delta\omega(t)$	The instantaneous phase difference
ω_e	The target phase difference
T	The period of the wave
Δt	The time offset between the two signals
S_i	Recorded signals
RS_i	Auxiliary signal
$EL(t)$	The electrical noise signal
$TR(t)$	Deformed triangular wave
$GAW(t)$	Gaussian white noise
β_1	Scaling factor
η_1/η_2	Weighting coefficients of the electrical noise
s	The slope of an oblique line associated with the deformed triangular wave
$ME(t)$	The mechanical vibration signal
$CO(t)$	Colored noise signals
μ_i	Weighting coefficients of the colored noise component

Chapter 4

S_{ij}	Sensor
Q_i	The MS signals received by the sensors (i= 1, 2, ... ,M)
L_S	The wave length of S-wave

T_i	The period of S-wave
P	The propagation distance difference
V_P	The velocity of P-wave
t_i	The triggering times of the sensors corresponding to signals Q_i
ω_e	The targeted phase difference
$\Delta\omega(t)$	The instantaneous phase difference

Chapter 5

$C_{S1s2}(\alpha)$	The normalized cross-correlation value of S_1 and S_2 at a time lag of α
t	Time lag
$A_T(t)$	The signaling component propagated along path T
τ_T	The traveltime difference along path T resulting from the different source locations
$R^{(t1,t2)}$	The correlation between two signals within the time interval from t_1 to t_2
ϵ	The time stretch factor
R_{max}	The maximum value of $R^{(t1,t2)}$
σ_τ	The time perturbation
ω_r	The mean square angular frequency of the recorded waveform
$\dot{s}(t)$	The time derivative of the waveform $s(t)$
v_p	The velocities of P-waves
v_s	The velocities of S-waves
${\delta_{cwi,ij}}^k$	The source separation between events i and j determined from the k-th sensor
$e_i = (x_i, y_i, z_i)$	The event location
N_{sensor}	The total number of sensors
n	The number of MS events considered for relative location
L	The residual of the objective function
t_s	The S-wave arrival time at the farthest sensor
t_b	The back-propagated traveltime from that sensor to the source
t_0	The original time of blasting event
r_{ij}^k	The arrival time difference between sensors i and j (where $i, j = 1, 2, \ldots n$, and n is the total number of sensors) for MS event k
(x, y)	The coordinates of the MS source
$\boldsymbol{B}_0$, $\boldsymbol{B}_1$, and $\boldsymbol{B}_3$	Lamé's parameter and the shear modulus
$\Delta\boldsymbol{b}$	A function of frequency $\boldsymbol{\omega}$ and quantifies the misfit between observed and simulated dispersion curves
$\boldsymbol{m} = (\boldsymbol{V}_s, \boldsymbol{V}_p, \boldsymbol{\rho})$	The model parameters

Chapter 6

$\boldsymbol{J}(\boldsymbol{m})$	The Jacobian matrix
W	Weighting matrix for the optimization

$\lvert\Delta b\rvert$	The absolute value of the residual vector Δb
$\lVert\lvert\Delta b\rvert\rVert_{inf}$	Largest element of $\lvert\Delta b\rvert$
$T^{\text{obs}}(r)$	The traveltimes measured at receivers r along the acquisition surface $\boldsymbol{\partial\Omega}$
g	The Fréchet matrix
$\boldsymbol{H}$	The Hessian matrix
$\boldsymbol{t(r)}$	The traveltime field satisfying the eikonal equation at receiver $\boldsymbol{r}$
$\boldsymbol{\lambda}(\boldsymbol{x})$	The adjoint-state parameter (Lagrangian multiplier) that imposes the constraint $\boldsymbol{t}(\boldsymbol{r}) = \boldsymbol{T^{obs}}$ at the optimal point
α	The step length
P_1, P_2, and P_3	Points selected to form the initial simplex triangle
y_1, y_2, and y_3	The three objective function values corresponding to points P_1, P_2, and P_3
y_{min}, y_{mid}, and y_{max}	The minimum, middle, and maximum values among y_1, y_2, and y_3
P^*, P^{**}, and P^{***}	The reflection, expansion, and contraction points

Chapter 7

E	Elastic modulus
μ	Poisson's ratio
ρ	Density of the rock
E_c	The current elastic modulus
E_0	The initial elastic modulus

Chapter 8

t	The MS wave traveltime
τ	The triggering time
d	The epicentral distance
α	A spatial searching radius
β	A density threshold
R_{dmax}	Distance threshold
Num_{smax} and Num_{smin}	Sensor selection criteria utilized for every source pair
Δs_m	The length of segments
M_{ik}	The number of segments
ω_{mn}	The interpolation weight of segments
$(T_k^{\,i}-T_k^{\,j})^{\text{obs}}$	The observing time difference
$(T_k^{\,i}-T_k^{\,j})^{\text{cal}}$	The calculating time difference
$\mathbf{A}$	The partial derivative matrix related to the hypocenter and origin time
$\mathbf{\Delta X}$	The perturbation vector for earthquake location and origin time
$\mathbf{C}$	The model derivative matrix associated with the slowness model

ΔM	The vector of slowness perturbations
ΔT	The misfit of arrival time
S	The station correction vector

Chapter 1
Introduction

1.1 Prologue

With the accelerated advancement of China's strategic initiatives "Deep Earth, Deep Sea, and Deep Space," the scale and number of deep underground engineering projects have grown substantially. In particular, tunnel infrastructures at kilometer-scale depths have become increasingly common, operating under complex geological conditions and high in situ stress environments (Liu et al. 2023). Rockburst, as one of the most prevalent and destructive geological hazards in deep tunneling, has demonstrated a marked escalation in both occurrence frequency and intensity, with its damaging consequences becoming increasingly severe (Zhou et al. 2018; Xiao et al. 2024, 2025). The underlying mechanism of rockburst is fundamentally attributed to the progressive accumulation of microfractures within the rock mass, culminating in a critical transition from distributed damage to catastrophic failure. During this process, the elastic strain energy stored in the surrounding rock is released instantaneously and violently, leading to the sudden ejection of rock fragments from the tunnel perimeter (Ren et al. 2023). Such dynamic failure not only causes extensive damage to excavation equipment and poses significant threats to personnel safety but also results in substantial disruption to construction schedules. In extreme cases, the occurrence of intense rockbursts can critically undermine the feasibility and long-term stability of the entire engineering project, thereby becoming a decisive factor in project success or failure (Zhou et al. 2023). Consequently, a comprehensive understanding of rockburst mechanisms and the development of effective prediction and mitigation strategies have become urgent research priorities in deep rock engineering.

Microseismic (MS) monitoring is widely recognized as the most direct and effective approach for rockburst monitoring and early warning in tunnels and has been extensively employed across a diverse range of tunnel and cave group engineering in recent years (Qiu and Zhou 2023; Wang et al. 2023; Tang et al. 2023). This technique enables real-time detection of elastic waves emitted by microfractures within the surrounding rock, thereby providing direct insight into the progressive

© Tongji University Press Co., Ltd 2026

Y. Xiao et al., *Near-Field Microseismic Data Processing in Deep Tunnels*,
https://doi.org/10.1007/978-981-92-3208-6_1

degradation and fracturing processes of the rock mass (Xue et al. 2021). Through a systematic sequence of MS signal processing steps, including event identification, phase picking, source localization, and inversion of source parameters, critical information regarding the rock fracturing process during rockburst incubation can be retrieved. Such information typically includes the occurrence time and spatial coordinates of rupture events, MS energy release, apparent stress, source radius, and moment magnitude, among other seismological attributes (Warpinski and Wolhart 2016). By integrating these MS source characteristics with site-specific geological conditions, the mechanical and deformation behavior of the surrounding rock, and the evolutionary trends of multiparameter MS indicators, it becomes possible to establish a dynamic and adaptive early warning framework for tunnel rockbursts. This integrated methodology not only enhances the timeliness and reliability of warning issuance but also supports the implementation of targeted mitigation measures under varying excavation stages and stress conditions (Qian et al. 2025; Xiao et al. 2026). Thus, MS monitoring has evolved into an indispensable component of intelligent risk management systems in deep tunneling engineering.

1.2 Book Outline

This book focuses on the investigation of MS data processing methodologies and proposes a series of innovative processing techniques. These methods are based on raw MS data and encompass signal recognition, P- and S-wave arrival time picking, source localization, velocity field inversion, and surrounding rock stability analysis. The ultimate objective of these approaches is to contribute to more accurate and expeditious rockburst early warning as well as the detection of the internal structure of the surrounding rock.

Chapter 2—Automatic rock fracture signal identification method based on tunnel MS data

This chapter introduces an innovative method for the identification of rock fracture signals based on waveform attributes. This technique emulates and digitizes the conventional manual interpretation of MS signals. The waveform recognition procedure is streamlined into three fully automated stages. The proposed method was evaluated across multiple tunnels excavated by a tunnel boring machine (TBM) (Jinping Tunnel) and drilling-and-blasting methods (Bayu Tunnel). The recognition accuracies for rock fracture signals reached 95.64% and 93.92%, with corresponding false-negative rates of 3.99% and 4.31%, respectively. Field applications in two distinct tunneling contexts confirm that this digital identification method is fully automated and operationally viable for MS monitoring in tunnel environments.

Chapter 3—P-wave arrival time picking method based on low signal-to-noise ratio (SNR) data without noise filtering

Due to the noisy construction environment, MS signals in tunnels often exhibit low signal-to-noise ratios (SNRs), making low-energy events difficult to distinguish from background noise using conventional filters. This study proposes a novel method called instantaneous phase difference intensity of P-waves (IPDI_P) to identify P-wave arrival times without noise filtering. The method requires that the propagation path difference between the target and auxiliary signals be within one P-wave wavelength, a condition typically met in tunnel monitoring. By constructing IPDI_P functions based on waveform correlation, the P-wave arrival is identified at the first peak approximating the global maximum. Noise characteristics (electrical and mechanical) were analyzed and simulated. Numerical experiments and a field case demonstrate that IPDI_P performs reliably, outperforming the short-term average to long-term average method and Akaike information criterion method, especially for SNRs below 30 dB. Mechanical vibration introduces greater picking errors than electrical noise. Principles for auxiliary signal selection are discussed.

Chapter 4—Determining S-wave arrival times for near-field MS data in tunnels

Accurate S-wave arrival picking is also essential for improving MS source location. The method of picking S-wave arrival times is similar to that of picking P-wave arrival times. This chapter proposes an indicator, which is the instantaneous phase difference intensity of S-waves (IPDI_S). This method is similar to IPDI_P. The IPDI_S peaks directly correspond to the initial S-wave break, enabling robust arrival time determination even when the arrival is obscured by the P-wave. Synthetic signal tests confirm that the method reliably identifies S-wave arrivals under P-wave interference. Field application in a deeply buried tunnel further demonstrates that IPDI_S can effectively pick S-wave arrivals that are indiscernible through manual inspection. As a result, the location accuracy of both rockburst and MS events is significantly enhanced.

Chapter 5—Absolute location method for MS event clusters in deep tunnels based on coda wave interference:

MS monitoring is essential for tunnel rockburst warning, with location accuracy being important to its predictive ability. Conventional methods require dense sensor arrays and precise velocity models but fail to account for the swarm behavior of MS events. This chapter proposes a coda wave interferometry (CWI)-based method incorporating event clustering, relative location, and absolute location. Relative locations are determined by sparse sensors and intersource velocity and then transformed to absolute coordinates via reference events. Numerical simulations yielded an average error of 1.46 m. Field tests showed errors of 2.95 m (blasting) and 3.65 m (rockburst). Compared to traditional methods, the proposed method produces more compact clustering and better aligns with physical phenomena, confirming its accuracy and robustness under sparse arrays and complex geological conditions.

Chapter 6—MS source location in deep buried tunnels using the updated inverted heterogeneous velocity:

This chapter presents a real-time velocity inversion method utilizing drilling and blasting data, enabling the construction of updated heterogeneous velocity models throughout progressive excavation stages. An optimized L2-norm double-difference algorithm is developed for MS source location based on these models. Numerical experiments confirm substantial improvements in location accuracy. Field validation at two deep tunnels in the Jinping, China Underground Laboratory, using multiple blasting sources with known coordinates further demonstrates the effectiveness and practicality of the proposed method.

Chapter 7—Analyzing the stability of rock surrounding deep cross-tunnels using a dynamic velocity field:

This chapter presents a dynamic stability assessment method for deeply buried cross-tunnels, integrating P-wave velocity inversion and rock damage evaluation. P-wave velocity is obtained via adjoint-state traveltime tomography (AST) following S-wave inversion from Rayleigh waves. The updated velocity field enables progressive calibration of the elastic modulus, while rock failure is simulated in CASRock using the equivalent plastic shear strain and local energy release rate. Applied to a case project in China, the method reveals that yield zones concentrate near sidewalls and exhibit a high–low–high–low pattern with distance from the adjacent tunnel. This work enhances the understanding of deep cross-tunnel construction.

Chapter 8—Stratified double-difference dynamic tomography of a 3D velocity field for a deep buried large underground cavern:

Accurate velocity models are essential for interpreting geological structures, excavation damage, and stress evolution in deep engineering. However, conventional time-lapse tomography is impractical for large-scale projects, and ill-conditioned sensor networks often hinder the convergence of passive inversion methods. To address these limitations, this chapter proposes a stratified double-difference (SDD) tomography method for three-dimensional dynamic velocity field inversion by MS events. The method resolves unknown parameters via double-difference analysis of traveltime residuals and employs a stratification strategy to mitigate adverse sensor–source geometry effects. Validated on checkerboard models with and without excavation zones at the Jinping underground laboratory, the approach achieves a resolution of approximately 50 m and cross-correlation coefficients of 0.84 and 0.79 with the true models. Dynamic inversion during excavation yields a cross-correlation of 0.71. These results demonstrate the robustness and applicability of the SDD method for dynamic velocity inversion in deep, large-scale underground caverns.

References

Liu L, Li S, Xiao Y et al (2023) An inverted heterogeneous velocity model for microseismic source location in deep buried tunnels. Rock Mech Rock Eng 56(7):4855–4880

Qian Y, Liu T, Zhai C, Wen H, Zhang Y, Zheng M, Gan X (2025) Real-time monitoring and analysis of hydraulic fracturing in surface well using microseismic technology: case insights and methodological advances. International Journal of. Mining Sci Technol

Qiu Y, Zhou J (2023) Short-term rockburst prediction in underground project: insights from an explainable and interpretable ensemble learning model. Acta Geotechnica 18(12):6655–6685

Ren F, Zhu C, He M, Shang J, Feng G, Bai J (2023) Characteristics and precursor of static and dynamic triggered rockburst: insight from multifractal. Rock Mech Rock Eng 56(3):1945–1967

Tang S, Li J, Tang L, Zhang L (2023) Microseismic monitoring and experimental study on rockburst in water-rich area of tunnel. Tunnell Underground Space Technol 141:105366

Wang K, Ma K, Tang CA, Liu H, Wang X, Li Q (2023) Study on deep mining-induced strata behavior based on the evolutional laws of multiple indices from microseismic monitoring. Rock Mecha Rock Eng 56(9):6481–6501

Warpinski NR, Wolhart S (2016) A validation assessment of microseismic monitoring. In: SPE hydraulic fracturing technology conference and exhibition. SPE, p (pD021S004R001)

Xiao Y, Chen S, Feng G et al (2025) Picking P-wave arrival times from microseismic signals in tunnels with low signal-to-noise ratios without noise filtering. Soil Dyn Earthquake Eng 192:109288

Xiao Y, Chen S, Liu L et al (2024) Determining S-wave arrival times for microseismic monitoring in tunnels based on instantaneous phase difference of sensors. Rock Mech Rock Eng 57(3):1781–1797

Xiao Y, Chen S, Liu L et al (2026) A stratified double-difference dynamic tomography of 3d velocity field for deep-buried large underground cavern: Y. Xiao et al. Rock Mech Rock Eng:1–24

Xue R, Liang Z, Xu N (2021) Rockburst prediction and analysis of activity characteristics within surrounding rock based on microseismic monitoring and numerical simulation. Int J Rock Mech Mining Sci 142:104750

Zhou J, Li X, Mitri HS (2018) Evaluation method of rockburst: state-of-the-art literature review. Tunnell Underground Space Technol 81:632–659

Zhou J, Zhang Y, Li C, He H, Li X (2023) Rockburst prediction and prevention in underground space excavation. Underground Space

Chapter 2
Automatic Rock Fracture Signal Identification Method Based on Tunnel MS Data

2.1 Prologue

Because of the inherently complex conditions characterizing underground construction environments, MS sensors can capture not only rock fracture signals but also waveform disturbances generated by various sources of construction activity, including vehicles, drilling, and other mechanized equipment. The accurate identification of rock fracture signals from background interference represents an important task but remains highly time-consuming when performed manually (He et al. 2023; Xiao et al. 2024, 2025). Thus, the development of automated identification methodologies for rock fracture events has increasingly become a shared research priority and engineering necessity (Xiao et al. 2026).

Current effective identification methods have two main categories: deep learning-based techniques and signal feature-driven methods.

(1) Deep learning models have been widely applied to the automatic recognition of rock fracture signals, such as convolutional neural networks (Dhanuka et al. 2023), long short-term memory networks (Karne et al. 2023), and artificial neural networks (Feng et al. 2023). However, high-performing architectures such as TF-CoV (Peng et al. 2020) and ResSCA (Shu et al. 2022) demand large-scale training sets consisting of 37,000 and 18,000 manually processed waveform samples, respectively. The effectiveness of such models heavily depends on the accuracy of manual dataset annotation (Ma et al. 2015).

(2) Signal feature-based methods emphasize the analysis of frequency-domain or time-frequency signal characteristics. These methods rely on advanced signal decomposition and reconstruction techniques, including empirical mode decomposition (Tang et al. 2021) and singular value decomposition (Bi et al. 2021). For example, Zhang et al. (2020) and Huang et al. (2018) applied empirical mode decomposition and local mean decomposition, respectively, for signal classification tasks. However, inherent drawbacks of signal processing algorithms restrict identification precision. Empirical mode decomposition, in particular, suffers

© Tongji University Press Co., Ltd 2026

Y. Xiao et al., *Near-Field Microseismic Data Processing in Deep Tunnels*,
https://doi.org/10.1007/978-981-92-3208-6_2

from mode mixing and endpoint effects (Malovichko 2012), which considerably degrade the reliability of automated signal identification.

In tunnel MS monitoring, manual identification based on visual inspection remains the prevailing approach due to both the scarcity of high-quality labeled datasets and the inherent limitations of signal processing algorithms. This chapter introduces a digitalization strategy for rock fracture signals that emulates the manual identification workflow. The digitization process comprises three sequential steps, each defined according to distinct characteristics of rock fracture signals. The proposed method and associated threshold parameters are initially validated using data from the Jinping Tunnel (excavated by a TBM). Subsequently, the same parameter configuration is applied to the Bayu Tunnel (excavated via the drilling and blasting method) to assess its adaptability across different construction environments.

2.2 Methodology

2.2.1 Manual Identification Process Review

Manual identification of rock fracture signals typically follows a three-step procedure based on waveform characteristics. The first step assesses the presence of distinct pulses: signals with clear pulses (Fig. 2.1a) proceed, while chaotic waveforms lacking pulse structure (Fig. 2.1b) are discarded. The second step evaluates the number of pulses within the signal. In tunnel environments, rock fracture events generally appear as a single pulse over a short time interval (Xiao et al. 2016). Multiple pulses suggest interference from construction activities, as shown in Fig. 2.1d. The final step classifies the signal type. Rock fracture signals are distinguished by a brief pulse with early peak amplitude and rapid attenuation (Fig. 2.1e). Waveforms that deviate from this pattern, such as Fig. 2.1f, are not attributed to fracturing events.

2.2.2 Digitalization Method

To simulate manual observation, a digitalization identification method is presented. This method consists of three steps, which are shown in Fig. 2.1g.

(1) Step 1: Normalization and K-means

A trace signal A_i to be determined is input. i is the sampling point ($i = 1,2, \ldots, I$). The absolute value of A_i is normalized.

$$A_i^n = \frac{|A_i|}{\max(|A_i|)} \tag{2.1}$$

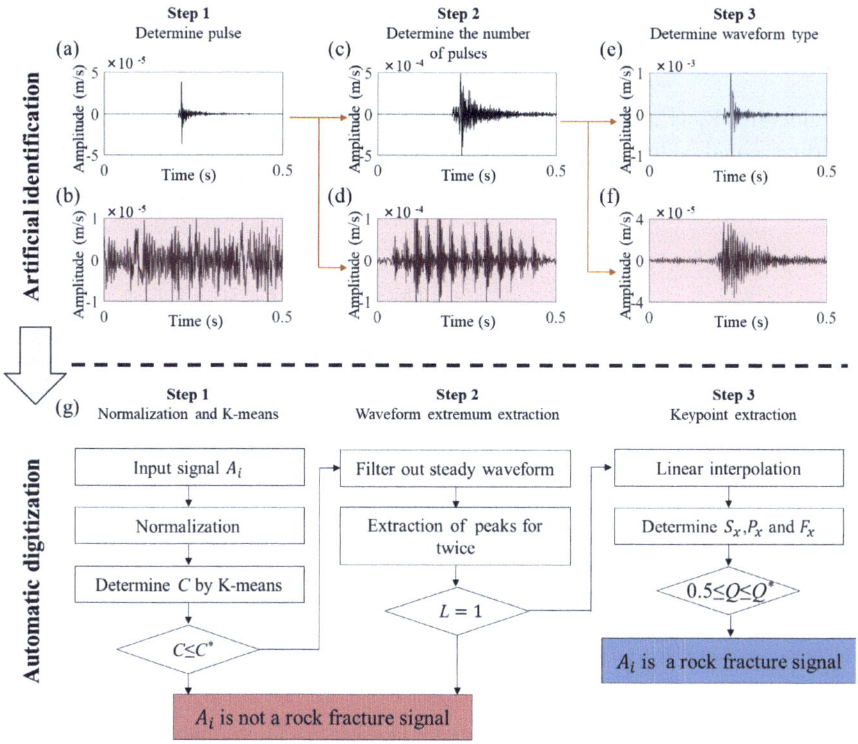

Fig. 2.1 Digitalization of rock fracture signal identification. **a–f** are the traditional manual identification methods for rock fracture signals. **a** Pulse-like signal. **b** Signal with no pulse. **c** Signal with only one pulse. **d** Signal with multiple pulses. **e** Rock fracture signal. **f** Other signals. **g** Workflow of the digitalization method

where "max" represents the operations for obtaining the maximum value. Then, the K-means method is introduced to compute C (cluster center). The detailed calculation process of this method can be found in Krishna and Narasimha (1999).

When C is less than the given threshold C^*, proceed to step 2; otherwise, A_i is not a rock fracture signal.

(2) Step 2: Waveform extremum extraction

The waveform is modified according to the following formula:

$$B = \{B_i\} = \begin{cases} A_i^n, A_i^n > 2 \times C \\ 0, A_i^n \leq 2 \times C \end{cases} \tag{2.2}$$

where B_i represents the modified waveform. Then, all the points that satisfy the given conditions are labeled y_j^1, where j=1, 2, …, J, constituting the set $Y_1 = \left\{ y_j^1 \right\}$.

$$\begin{cases} B_i - B_{i-1} > 0 \\ B_i - B_{i+1} > 0 \end{cases} \tag{2.3}$$

Evidently, Y_1 is a subset of B, indicating that each element in Y_1 has a corresponding equal value in B. The positions are labeled x_j^1, constituting the set $X_1 = \left\{x_j^1\right\}$. Next, all the points satisfying the following conditions are extracted:

$$\begin{cases} y_j^1 - y_{j-1}^1 > 0 \\ y_j^1 - y_{j+1}^1 > 0 \end{cases}, j = 1, 2, \ldots, J \tag{2.4}$$

These points are then labeled y_k^2, where k=1, 2, …, K, constituting the set $Y_2 = \left\{y_k^2\right\}$ The positions of y_k^2, in B are labeled x_k^2, constituting the set $X_2 = \left\{x_k^2\right\}$.

y_k^2, with values exceeding 0.5 are screened, and a check is performed to determine whether k is continuous. If k is continuous, the signal has only one pulse waveform ($L = 1$). Then, move to step 3. Otherwise, A_i is not a rock fracture signal.

(3) Step 3: Keypoint extraction

i, A_i, x_j^1, x_k^2, y_j^1, and y_k^2, have the following relationships:

$$Y_2 \subseteq Y_1 \subseteq B, X_2 \subseteq X_1 \subseteq \{1, 2, \ldots, I\} \tag{2.5}$$

The positions of x_k^2 can be identified and labeled p_k. Then, linear interpolation is performed between $y_{p_k}^1$ and $y_{p_{k+1}}^1$, and $p_{k+1} - p_{k-1}$ points are inserted. The positions of the inserted points are $x_{p_{k+1}}^1, x_{p_{k+2}}^1, \ldots, x_{p_{k+1}-1}^1$.

$$\begin{cases} y_{p_k+1}^{2*} = y_{p_k}^1 + \frac{y_{p_{k+1}}^1 - y_{p_k}^1}{x_{p_{k+1}}^1 - x_{p_k}^1} \times \left(x_{p_k+1}^1 - x_{p_k}^1\right) \\ y_{p_k+2}^{2*} = y_{p_k}^1 + \frac{y_{p_{k+1}}^1 - y_{p_k}^1}{x_{p_{k+1}}^1 - x_{p_k}^1} \times \left(x_{p_k+2}^1 - x_{p_k}^1\right) \\ \cdots \\ y_{p_{k+1}-1}^{2*} = y_{p_k}^1 + \frac{y_{p_{k+1}}^1 - y_{p_k}^1}{x_{p_{k+1}}^1 - x_{p_k}^1} \times \left(x_{p_{k+1}-1}^1 - x_{p_k}^1\right) \end{cases} \tag{2.6}$$

where $y_{p_{k+1}}^{2*}, y_{p_{k+2}}^{2*}$, and $y_{p_{k+1}-1}^{2*}$ are the amplitude values inserted at positions $x_{p_{k+1}}^1, x_{p_{k+2}}^1$, and $x_{p_{k+1}-1}^1$, respectively. Finally, a new set Y_{2*} is labeled, consisting of the inserted points and y_k^2,. The corresponding sampling point is extracted as P_x when Y_{2*} equals 1. By the order of the sampling points, backward and forward searches are performed from P_x to find the first points with an amplitude smaller than 0.2. These points are labeled S_x and F_x, respectively. Then, the ratio Q is calculated.

$$Q = \frac{F_x - P_x}{P_x - S_x} \tag{2.7}$$

A satisfactory $0.5 \leq Q \leq Q^*$ indicates that the signal is a rock fracture signal.

2.2.3 Threshold Determination

The proposed digitalization method requires the determination of two thresholds, denoted as C^* and Q^*, based on manually annotated field data. For each signal, the parameters C and Q are computed following the procedure in Fig. 2.1g prior to threshold selection. To identify the optimal threshold pair (C^*, Q^*), the accuracy rate (ACC) and miss rate (MIS) are adopted as evaluation criteria.

$$\mathrm{ACC} = \frac{N_{\mathrm{m\&w}}}{N_{\mathrm{w}}} \times 100\% \tag{2.8}$$

$$\mathrm{MIS} = \frac{N_{\mathrm{o}}}{N_{\mathrm{m}}} \times 100\% \tag{2.9}$$

where $N_{\mathrm{m\&w}}$, N_{w}, N_{o}, and N_{m} represent the number of rock fracture signals identified in different ways. N_{m}, N_{w}, $N_{\mathrm{m\&w}}$, N_{o} represent manual identification, digital identification, both manual and digital identification, and signals omitted by the digitalization method, respectively. Then, a suitability rate α is defined and used to evaluate which threshold pair is optimal.

$$\alpha = 0.5 \times \mathrm{ACC} + 0.5 \times (1 - \mathrm{MIS}) \tag{2.10}$$

The optimal threshold pair corresponds to the maximum of α.

2.3 Field Case

2.3.1 Field Case Background

The proposed digitalization method was implemented and evaluated using MS monitoring data from the Jinping Tunnel, situated in Sichuan Province, China. Detailed geological information for this tunnel is available in Xu et al. (2016). The configuration of the MS monitoring network is shown in Fig. 2.2a. Two sensor groups, designated Group 1 and Group 2, were installed at 25 m intervals along the tunnel. As the TBM progressed, sensors in Group 1 were relocated to the position of Group 3. This iterative repositioning enabled the sensor array to advance continuously along the excavation direction, thereby supporting real-time dynamic monitoring of MS events.

The large number of signals acquired by MS systems necessitates rapid and accurate discrimination of rock fracture events to enhance rockburst warning efficiency. Thus, the development of automated and reliable identification methods is imperative.

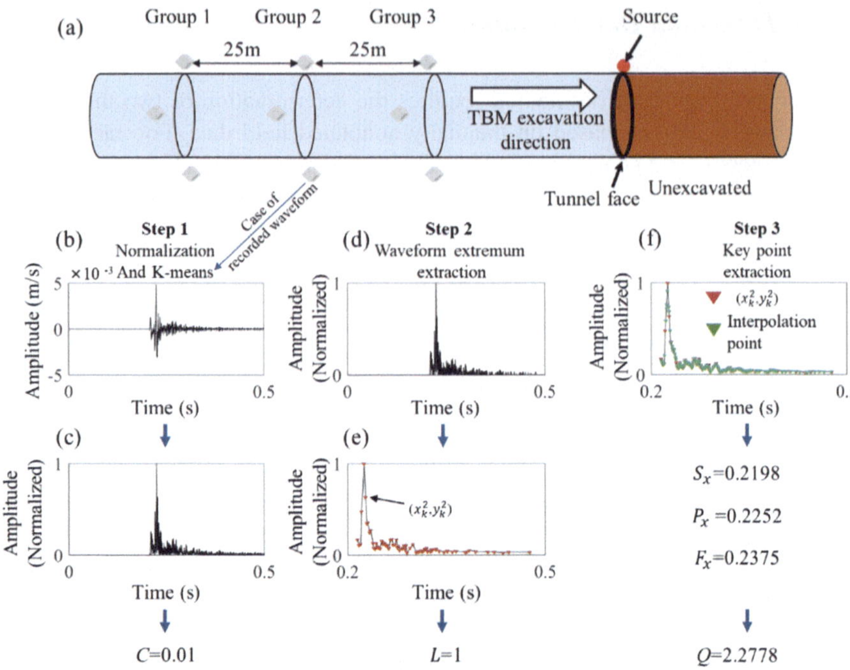

Fig. 2.2 Case study in the Jinping Tunnel. **a** Sensor arrangement. **b** Original signal case. **c** A_i^n. **d** B_i. **e** $\left(x_k^2, x_k^2\right)$. **f** Linear interpolation results

2.3.2 Threshold in the Jinping Tunnel

Thresholds were established using signals recorded from June 15 to 17, 2010, totaling 336 events, of which 95 were manually identified as rock fractures. Parameters Q and C were computed for all signals. Q could be extracted from only 31% of the other signals. Inspection of the parameter distributions (Fig. 2.3a) indicates that rock fracture signals cluster within the region defined by $C < 0.05$ and $Q < 5$. The coefficient α, derived from Eq. 2.10, reaches its maximum at the threshold pair ($C^* = 0.05$, $Q^* = 5$), as shown in Fig. 2.3b. This threshold combination was subsequently validated using a four-month dataset (June to September), which consistently yielded the same optimal values ($C^* = 0.05$, $Q^* = 5$) across the entire period.

2.3.3 Case Analysis

(1) Single waveform digitalization example

A representative waveform (A_i, i = 1–3000) sampled at 6000 Hz is illustrated in Fig. 2.2b. After normalization (Fig. 2.2c), K-means clustering yields a C value of

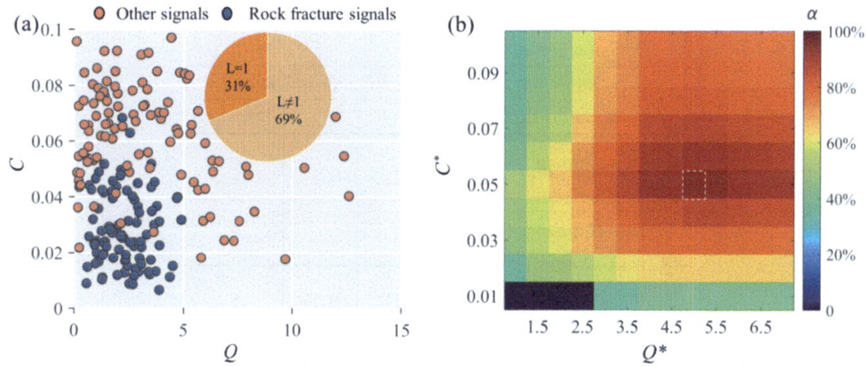

Fig. 2.3 Determination of the threshold pair in the Jinping tunnel. **a** Parameter distributions. **b** Suitability rate α under different threshold pairs

Table 2.1 Amplitude exceeding 0.5 in Fig. 2.2e

	Point 1	Point 2	Point 3
$y_{k'}^2$	0.51	1	0.63
k	4	5	6

0.01. The waveform is subsequently corrected via Eq. 2.2 to produce B_i (Fig. 2.2d), from which peak points are extracted following Eqs. 2.3–2.4 (Fig. 2.2e). Consistent with Step 2, all points in B_i with amplitudes exceeding 0.5, along with their corresponding indices k, are listed in Table 2.1. The trend of these indices indicates that $L = 1$.

Q is subsequently derived through linear interpolation applied to y_k^2, following Eq. 2.6. For example, points y_4^2 and y_5^2 correspond to indices $j = 15$ and 20 in y_j^1, respectively. Interpolated values are then introduced at each integer index between these two positions ($j = 16$–19). After interpolation is completed (Fig. 2.2f), the points (S_x, S_y) and (F_x, F_y) are identified in the dataset.

Utilizing point (P_x, P_y) as the central reference, where $P_x = 0.2252$, a search is performed for an initial value smaller than 0.2 both in the forward and backward directions in Fig. 2.2f. The associated index coordinates are identified as $S_x = 0.2198$ and $F_x = 0.2375$. Consequently, the parameter Q is computed as 2.2778.

(2) Digitalization identification examples

Utilizing ten representative signals as illustrative instances (Table 2.2), a systematic exposition of the procedure for identifying signals characteristic of rock failure is presented.

Initially, C associated with the ten signals is determined. Following the implementation of a threshold criterion established at 0.05, signals 1, 2, 5, 7, and 8 are excluded from further examination. An evaluation is conducted to determine whether the parameter L attains a value of 1 within the set of retained signals. The findings of this assessment lead to the exclusion of signal 6 from continued analysis. The Q

Table 2.2 Typical signal identification case (if the signal is identified as a rock fracture signal, it is marked as "√" in the "Decision" column)

No.	Time-waveform	Digitalization identification			
		C	*L*=1	*Q*	Decision
1		0.12	/	/	×
2	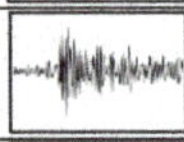	0.09	/	/	×
3		0.04	=1	4.56	√
4	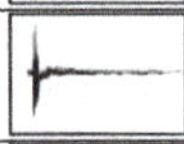	0.02	=1	1.46	√
5		0.10	/	/	×
6	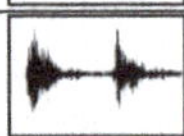	0.04	≠1	/	×
7		0.10	/	/	×
8		0.06	/	/	×
9	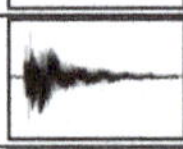	0.05	=1	8.26	×
10	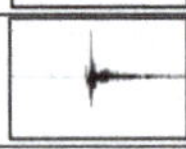	0.02	=1	2.22	√

values pertaining to the remaining signals are then assessed to verify their conformity with the designated range of 0.5–5. The outcomes confirm that signals 3, 4, and 10 fulfill all stipulated requirements and are consequently classified as rock fracture signals.

2.3.4 Performance Assessment

To further evaluate the digitalization method, its performance was assessed using field data recorded from June 18 to July 14. During this period, 8,287 MS signals were captured, of which 1243 (15.00%) were identified as rock fracture events. Data from June 27 to July 1 were unavailable due to system maintenance.

The daily variations in the ACC and MIS rates varied between 92.31–98.40% and 1.67–7.69%, respectively (Fig. 2.4), with mean values of 95.64% and 3.99%. The temporal stability of both metrics throughout the monitoring period indicates that the proposed digitalization method consistently achieves high identification accuracy and a low false-negative rate.

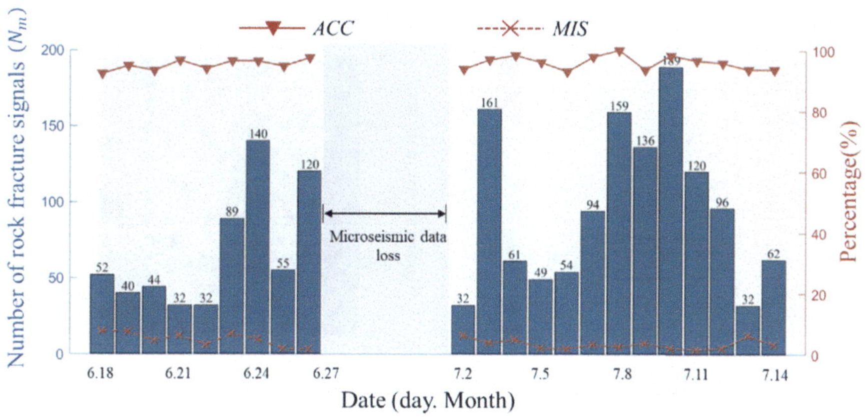

Fig 2.4 Identified results of rock fracture signals using the field data collected from 18th June to 14th July in the Jinping Tunnel for MS monitoring

2.4 Discussion

2.4.1 Factors That Determine Thresholds

Threshold determination constitutes a critical component of the proposed digitalization method. To examine the two influencing factors, the amount of MS training data and the proportion of rock fracture signals, multiple training sets were constructed with varying sizes (100 to 500 signals) and rock fracture signal percentages (10% to 90%). The thresholds derived from each dataset were subsequently applied to rock fracture signal identification using MS monitoring data collected from the Jinping Tunnel between June 18 and July 14. The resulting ACC and MIS values for each configuration are presented in Fig. 2.5.

The results indicate that ACC and MIS remain consistent across different training set sizes, with reasonable identification performance attainable even from relatively small datasets (e.g., 200 signals). In contrast, the proportion of rock fracture signals substantially influences both metrics: a lower percentage elevates both accuracy (higher ACC) and the false-negative rate (higher MIS). This introduces a trade-off between ACC and MIS when optimizing the fracture signal percentage. Considering both metrics, a training set comprising 200–500 signals with a rock fracture signal proportion of 15–35% is recommended for the digitalization method (indicated by the white box in Fig. 2.5).

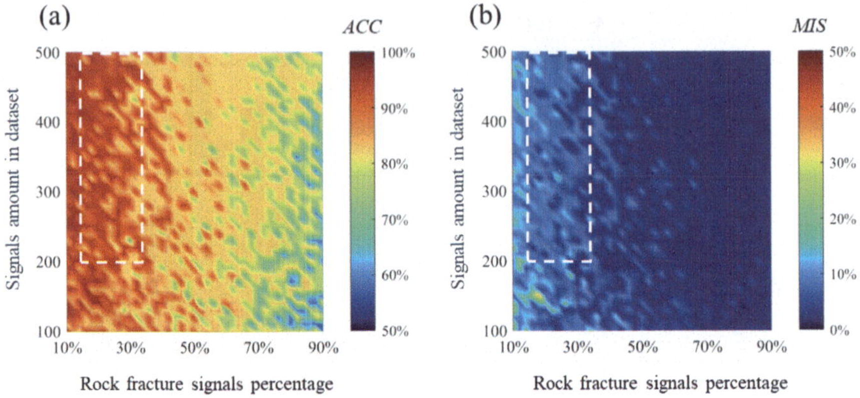

$$\text{Rock fracture signals percentage} = \frac{\text{The number of rock fracture signals}}{\text{Signals amount in dataset}} \times 100\%$$

Fig. 2.5 Identification performance using different datasets. **a** ACC and **b** MIS results

2.4.2 Further Application in a Drill and Blast Tunnel

The proposed method was also applied to the Bayu Tunnel, which was excavated using the drilling and blasting method. The MS monitoring in this tunnel was similar to that in the Jinping Tunnel, and detailed geological information is available in Hu et al. (2020). Threshold determination was performed using MS data recorded from October 1 to October 3, 2018. The coefficient α attained a maximum value of 95.32% at $C^* = 0.06$ and $Q^* = 5$, identifying this pair as the optimal threshold for rock fracture signal recognition in the Bayu Tunnel.

The method was subsequently validated on a dataset comprising 7,043 MS signals recorded from October 8 to October 31, 2018, of which 1,348 were manually identified as rock fracture events. The resulting ACC values ranged from 91.75 to 95.89%, with a mean of 93.92%, while MIS varied between 1.99 and 7.98%, averaging 4.31%. These results demonstrate that the digitalization method, initially developed for a TBM-excavated tunnel, performs effectively in a drilling and blasting context as well.

2.4.3 Strengths and Limits

The strengths and limitations of the proposed digitalization-based identification method are summarized as follows:

(1) Automated signal recognition technology can save human resources. Most automatic identification methods are compared to manual identification methods,

which suggests that the manual identification result is the ground truth. The digitalization method in this paper simulates a human to identify rock fracture signals in tunnel MS monitoring and perform manual identification consistently. This method achieves higher accuracy than other methods.

(2) The threshold exhibits temporal stability within the same project and remains effective across different tunnel excavation conditions, as demonstrated by the successful application in drilling and blasting tunnels. This indicates the adaptability of this method to varying construction conditions.

(3) The proportion of rock fracture signals in the training dataset requires careful consideration, as it significantly influences the trade-off between achieving high identification accuracy and maintaining a low false-negative rate.

2.5 Conclusion

This paper introduces a digitalization method that comprises three main steps: normalization and K-means calculation, waveform extremum extraction, and key point extraction. The method is fully automated and designed to replicate and quantify the manual process of identifying rock fracture signals in tunnel MS monitoring.

The proposed method was successfully implemented in two tunneling environments: the Jinping Tunnel (excavated by TBM) and the Bayu Tunnel (excavated by drilling and blasting). It yielded average accuracies of 95.64% and 93.92%, respectively, with corresponding false-negative rates of 3.99 and 4.31%.

Threshold determination plays a critical role in the method's performance and requires an appropriately selected training dataset. Optimal and stable thresholds can be achieved using a dataset comprising 200–500 MS signals, with rock fracture signals accounting for 15–35% of the total. The required training data volume is substantially lower than that typically demanded by deep learning-based methods. Moreover, the obtained thresholds maintained consistent performance across both TBM and drill-and-blast excavation conditions.

References

Bi X, Zhang C, He Y, Zhao X, Sun Y, Ma Y (2021) Explainable time–frequency convolutional neural network for microseismic waveform classification. Inf Sci 546:883–896

Dhanuka R, Singh JP, Tripathi A (2023) A comprehensive survey of deep learning techniques in protein function prediction. IEEE/ACM Transa Comput Biol Bioinf 20(3):2291–2301

Feng Q, Han L, Zhao B, Li Q (2023) Microseismic events recognition via joint deep clustering with residual shrinkage dense network. IEEE Trans Geosci Remote Sens 61:1–11

He Z, Jia M, Wang L (2023) UACNet: a universal automatic classification network for microseismic signals regardless of waveform size and sampling rate. Eng Appl Artificial Intell 126:107088

Hu L, Feng XT, Xiao YX, Wang R, Feng GL, Yao ZB, Zhang W (2020) Effects of structural planes on rockburst position with respect to tunnel cross-sections: a case study involving a railway tunnel in China. Bull Eng Geol the Environ 79(2):1061–1081

Huang L, Li J, Hao H, Li X (2018) Micro-seismic event detection and location in underground mines by using Convolutional Neural Networks (CNN) and deep learning. Tunnell Underground Space Technol 81:265–276

Karne A, Karne R, Kumar MVK, Arunkumar A (2023) Convolutional neural networks for object detection and recognition. J Artif Intell Mach Learn Neural Netw 3(3):1–13

Krishna K, Murty MN (1999) Genetic K-means algorithm. IEEE Trans Syst Man Cybern Part B (Cybernetics) 29(3):433–439

Liu L, Li S, Xiao Y et al (2023) An inverted heterogeneous velocity model for microseismic source location in deep buried tunnels. Rock Mech Rock Eng 56(7):4855–4880

Ma J, Zhao G, Dong L, Chen G, Zhang C (2015) A comparison of mine seismic discriminators based on features of source parameters to waveform characteristics. Shock Vibr 1:919143

Malovichko D (2012) Discrimination of blasts in mine seismology. In: Deep mining 2012: proceedings of the sixth international seminar on deep and high stress mining, pp 161–172

Peng P, He Z, Wang L, Jiang Y (2020) Automatic classification of microseismic records in underground mining: a deep learning approach. IEEE Access 8:17863–17876

Shu L, Liu Z, Wang K, Zhu N, Yang J (2022) Characteristics and classification of microseismic signals in heading face of coal mine: implication for coal and gas outburst warning. Rock Mech Rock Eng 55(11):6905–6919

Tang S, Wang J, Tang C (2021) Identification of microseismic events in rock engineering by a convolutional neural network combined with an attention mechanism. Rock Mech Rock Eng 54(1):47–69

Xiao Y, Chen S, Feng G et al (2025) Picking P-wave arrival times from microseismic signals in tunnels with low signal-to-noise ratios without noise filtering. Soil Dyn Earthquake Eng 192:109288

Xiao Y, Chen S, Liu L et al (2024) Determining S-wave arrival times for microseismic monitoring in tunnels based on instantaneous phase difference of sensors. Rock mechanics and rock engineering 57(3):1781–1797

Xiao YX, Feng XT, Hudson JA, Chen BR, Feng GL, Liu JP (2016) ISRM suggested method for in situ microseismic monitoring of the fracturing process in rock masses. Rock Mech Rock Eng 49(1):343–369

Xiao Y, Chen S, Liu L, Li S, Chen B, Feng G (2026) Location for microseismic event clusters in deep tunnels based on coda wave interference. Tunnell Underground Space Technol 174:107690

Xu NW, Li TB, Dai F, Zhang R, Tang CA, Tang LX (2016) Microseismic monitoring of strainburst activities in deep tunnels at the Jinping II Hydropower Station, China. Rock Mech Rock Eng 49(3):981–1000

Zhang H, Ma C, Pazzi V, Li T, Casagli N (2020) Deep convolutional neural network for microseismic signal detection and classification. Pure Appl Geophys 177(12):5781–5797

Chapter 3
P-Wave Arrival Time Picking Method Based on Low Signal-to-Noise Ratio (SNR) Data Without Noise Filtering

3.1 Prologue

A critical factor in the precise location of MS events lies in the key to the successful application of MS monitoring, which in turn relies on precise picking of P- and S-wave arrival times (He et al. 2020; Xiao et al. 2024, 2025; Liu et al. 2023). The two most widely adopted techniques for P-wave arrival picking are the short-term average to long-term average (STA/LTA) method (Allen 1978) and the Akaike information criterion (AIC) method (Akaike 1974). The STA/LTA approach identifies arrivals by computing the ratio of signal averages within short and long moving windows, necessitating the specification of a threshold. Its accuracy is contingent upon the appropriate selection of feature functions, window lengths, and dynamic thresholds (Vaezi and Van der Baan 2015; Khalqillah et al. 2018; Wong et al. 2009; Sabbione and Velis 2013; Chen et al. 2019; Withers et al. 1998). In contrast, the AIC method determines arrival times by locating the first minimum in the instantaneous chaos level of the signal. Several variants have since emerged (Song and Toksöz 2011; Li et al. 2017), with autoregressive AR-AIC (Morita 1984) and vector autoregressive VAR-AIC (Maeda 1985) remaining among the most established and frequently employed. Despite their widespread use, both methods exhibit sensitivity to noise and perform poorly under low signal-to-noise ratio (SNR) conditions, frequently yielding inaccurate picks (Vaezi and Van der Baan 2015). To mitigate this limitation, numerous hybrid approaches integrating noise reduction have been developed, including DWT-STA/LTA (Li et al. 2016), k-means-STA/LTA (Akram et al. 2018), EMD-AIC (Li et al. 2017), and shearlet-AIC (Gong et al. 2019). In essence, effective noise filtering remains a prerequisite for reliable arrival picking in low-SNR MS signals.

In recent years, advances in deep learning have introduced new methods for signal analysis, with neural network-based techniques demonstrating considerable promise (Nejad and Jaksa 2017; Dhanuka et al. 2023; Feng et al. 2023). Architectures such as U-net (Zhang et al. 2021) and recurrent neural networks (Zheng et al. 2018) enable

© Tongji University Press Co., Ltd 2026

Y. Xiao et al., *Near-Field Microseismic Data Processing in Deep Tunnels*,
https://doi.org/10.1007/978-981-92-3208-6_3

P-wave arrival identification without the need for manually defined thresholds or criteria. However, because these models are typically trained on datasets comprising MS signals with clear, high-quality P-wave arrivals, their performance degrades when applied to signals with low SNRs. Consequently, even within deep learning frameworks, effective noise preprocessing remains essential for accurate arrival time determination.

MS signals recorded in tunnel environments are typically subject to severe noise contamination. As a result, existing methods for P-wave arrival picking generally necessitate noise filtering as a preliminary step. However, different noise sources often require different filtering strategies, and the filtering process itself may introduce additional errors in arrival time identification. Thus, there is a clear need for a method capable of directly determining P-wave arrival times from raw MS signals without prior noise suppression. This chapter aims to address this challenge.

The investigation begins with an analysis of source wave propagation paths, noise characteristics, and SNR distributions specific to tunnel MS signals. Based on these insights, a novel picking method grounded in the concept of instantaneous phase difference is proposed, which operates without the need for noise filtering. Numerical simulations are conducted to evaluate the performance of the proposed approach. Subsequently, the method is applied to locate MS events associated with an actual moderate-intensity rockburst case. Comparative analysis demonstrates that the proposed method outperforms traditional STA/LTA and AIC techniques in terms of picking accuracy and reliability.

3.2 MS Monitoring in Tunnels

3.2.1 Data from the Jinping Tunnel

MS sensors are typically installed in two rows behind the working face in a tunnel, as shown in Fig. 3.1. Although the exact placement may vary depending on whether TBM or drilling and blasting methods are used, the spacing between the two rows generally falls within a consistent range of 20 to 35 m.

The configuration adopted during excavation of the #3 TBM tunnel of the Jinping II hydropower station is shown in Fig. 3.1. Detailed descriptions of this 2465-m deep diversion tunnel project can be found in Xu et al. (2016). The MS data analyzed in this study were recorded in this tunnel between July 14 and July 21, 2009. During this period, 74 MS events resulting from rock fracturing generated a total of 436 rock-fracture signals. As excavation progressed from 10,743.84 to 10,645.92, the sensor arrays underwent relocation three times to maintain proximity to the advancing working face (Fig. 3.2a).

At 22:51 on July 21, 2009, a moderate-intensity rockburst (hereafter referred to as the "7/21" rockburst) occurred, creating an explosion pit approximately 0.2 to 0.5 m deep. The event affected a broad section of the tunnel, spanning from 10,670

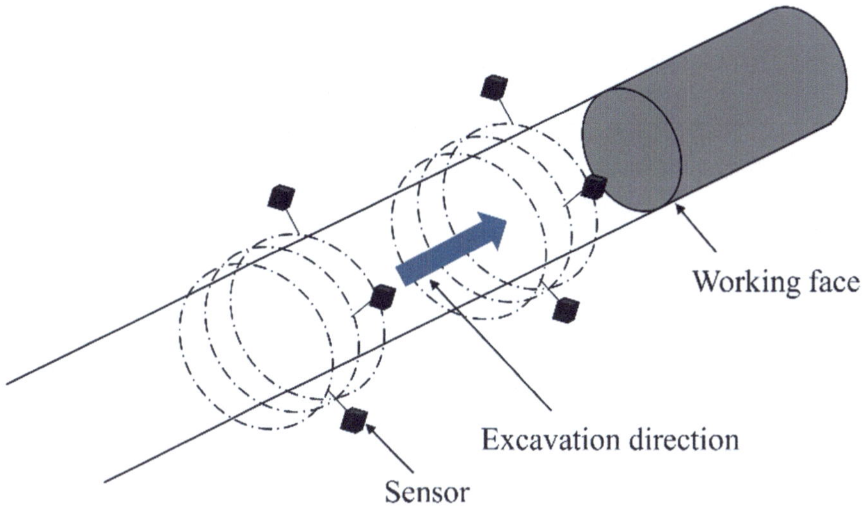

Fig. 3.1 The sensor layout

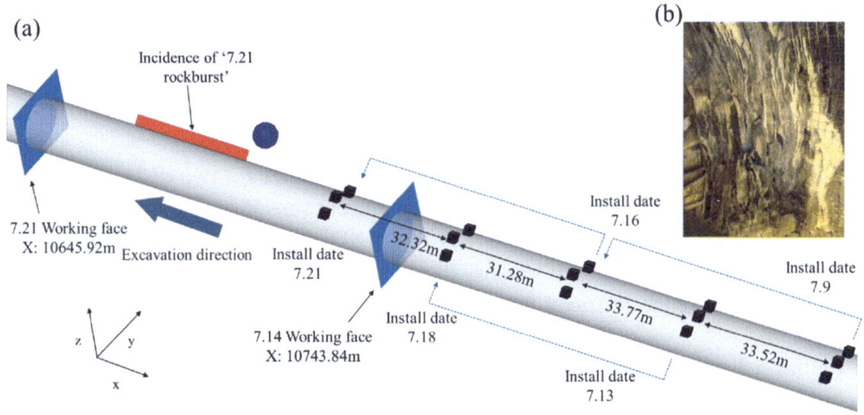

Fig. 3.2 The position of the rockburst

to 10,700 along the axis. The location of associated MS events indicated that the rockburst focus was situated at 10,682. The spatial distribution of the 7/21 rockburst and the resulting rock damage are presented in Fig. 3.2a, b, respectively.

3.2.2 *Source Wave Traveling Path Difference Versus Wavelength of the P-Wave*

Xiao et al. (2015) observed that when an MS event occurs in a tunnel, the dominant frequency component of the signal undergoes less than 10% attenuation as it propagates from the nearest sensor row to the adjacent row. This indicates that the frequency attenuation between sensor arrays is negligible. Consequently, an average frequency F_a is introduced to unify the dominant frequencies of signals originating from the same MS event, defined as follows:

$$F_a = \frac{1}{M-2}\left(\sum_{i=1}^{M} f_i - \max(f_i) - \min(f_i)\right) \tag{3.1}$$

where M is the number of MS signals associated with the event and f_i is the main frequency in the i-th MS signal. In tunnel environments, the propagation velocity of P-waves, denoted as V_p, is generally assumed to be constant and can be determined through velocity inversion analysis. As a result, the wavelengths of P-waves across different MS signals originating from the same event are expected to be comparable. This unified wavelength, L_p, can therefore be estimated using the following expression:

$$L_P = V_P / F_a \tag{3.2}$$

To facilitate comparison with the wavelength parameter L_p, two distance metrics are introduced: the shortest distance between different sensor rows (SDDSR) and the source wave traveling path difference (SWTPD). SDDSR is defined as the minimum spatial separation between two sensors located in adjacent rows, whereas SWTPD denotes the difference in travel path lengths from the seismic source to these two sensors.

Figure 3.3 presents the distribution of SDDSR relative to L_p for the 436 signals recorded in the Jinping tunnel. The results indicate that for the vast majority of cases (91.7%), the SDDSR/L_p ratio is less than 1. According to the triangle inequality theorem, the difference between any two sides of a triangle is always less than the third side. Consequently, the SWTPD between two MS signals cannot exceed the spatial distance between the sensors that recorded them. Thus, for most signal pairs, the SWTPD is smaller than the P-wave wavelength, satisfying the inequality SWTPD $\leq$ SDDSR $< L_p$.

3.2.3 *Characteristics of Noise Signals*

In tunnel environments, MS signals are primarily contaminated by noise originating from five sources: blasting, drilling, electrical equipment, mechanical vibration, and

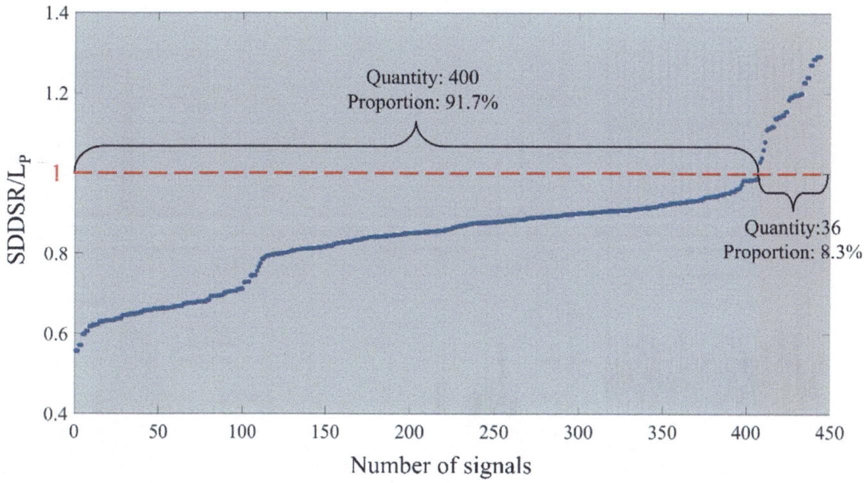

Fig. 3.3 Distribution of the SDDSR/L_p values measured for the MS signals

rock fracturing (Xiao et al. 2016). Signals generated by rock fracturing are generally unaffected by blasting or drilling noise, as these activities rarely coincide temporally. Blasting events are infrequent, typically occurring once per excavation cycle in drilling and blasting tunnels. Drilling signals are concentrated within short time intervals. In TBM tunneling, drilling is mainly performed during excavation pauses and seldom exceeds 20 minutes in duration.

In contrast, electrical noise and mechanical vibration can occur randomly at any time, and thus, their influence on MS signal quality is of particular concern. Representative waveforms, frequency spectra, and amplitude distributions for these two noise types are presented in Fig. 3.4. The characteristics of electrical and mechanical noise differ markedly. Electrical noise exhibits strong periodicity (Fig. 3.4a) and can be decomposed into discrete frequency components concentrated around specific values, with a dominant frequency of 50 Hz (Fig. 3.4b). Mechanical vibration, by comparison, is aperiodic and stochastic (Fig. 3.4d), comprising a broad frequency range of 200 to 800 Hz (Fig. 3.4e).

Despite these differences, the sampling amplitudes of both noise types conform to Gaussian distributions with expectations near zero. However, their variances diverge: the variance of electrical noise ranges from 0.1 to 0.2 times its normalized maximum amplitude, whereas that of mechanical noise falls between 0 and 0.1 times (Figs. 3.4c, f). It should be noted that variance is computed from the original signal amplitudes, while the mathematical expectation is derived from the amplitude distribution following statistical analysis.

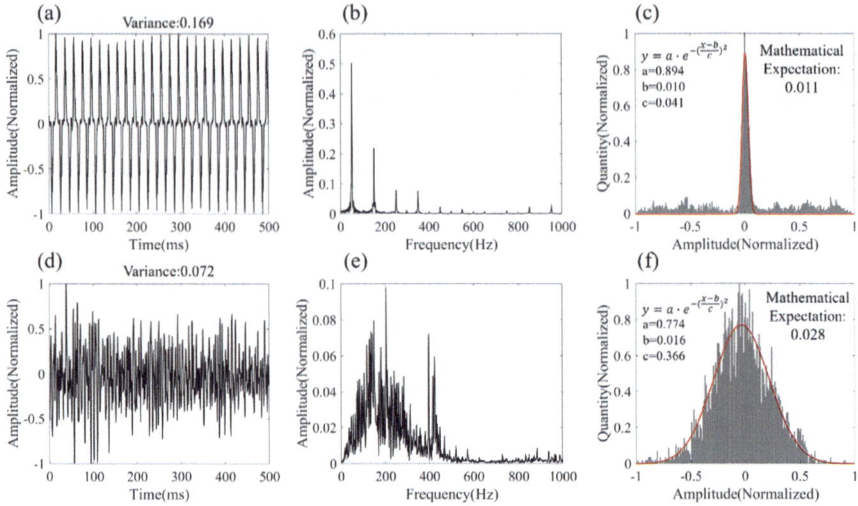

Fig. 3.4 The distinct characteristics of two types of noise signals. Electrical noise exhibits pronounced periodicity in the timedomain (**a**), can be represented by a limited number of dominant frequency components (**b**), and shows amplitude fluctuations centered near zero (**c**). In contrast, mechanical vibration signals are more random and aperiodic in nature (**d**), span a broad frequency range (**e**), and display larger amplitude variations (**f**)

3.2.4 SNR of Rock-Fracturing Signals

Given that completely separating a target signal from background noise is challenging, accurate calculation of the signal's SNR is often difficult. In this study, the SNR of rock fracturing signals is estimated by the approach proposed by Fard et al. (2020), as expressed in the following formula:

$$\mathrm{SNR} = \frac{\mathrm{RMS(MS)}}{\mathrm{RMS(Noise)}} \tag{3.3}$$

which essentially compares the root mean square (RMS) values of the signal over two time windows. Both windows are set to an identical duration of 50 ms. However, they serve different purposes: the MS window is aligned with the signal's peak amplitude, covering the interval from 15 ms before to 35 ms after the maximum point, while the noise window is temporally separated from the peak, positioned between 150 ms and 100 ms prior to it.

Figure 3.5a presents a series of rock fracturing signals with SNRs ranging from 5 to 45 dB. When the SNR is sufficiently high (specifically, no less than 30 dB), the P-wave arrival time can be reliably identified. However, as the SNR decreases to 25 dB or below, the arrival becomes increasingly obscured by background noise, rendering manual picking difficult and progressively more inaccurate with declining SNR. The SNR distribution for the 436 rock fracturing signals recorded in the Jinping tunnel, calculated using Eq. 3.3, is shown in Fig. 3.5b. Among these, 185 signals (42.4%) exhibit SNRs above 30 dB and can be effectively utilized for P-wave arrival identification without prior noise filtering. In contrast, 158 signals (36.2%) have SNRs below 26 dB, placing them in a range where accurate manual picking is challenging. It is important to note that the SNR of the P-wave itself is generally lower than that of the entire signal (Eisner et al. 2009; Liu et al. 2020). For example, a signal with an overall SNR of 30 dB typically corresponds to a P-wave SNR in the range of 16–24 dB. Further analysis can be conducted on the SNRs of signals belonging to the same MS event. For each event, the average $\mathrm{SNR}_{\mathrm{Average}}$ and standard deviation $\mathrm{SNR}_{\mathrm{standard}}$ can be computed using the following expressions:

$$\mathrm{SNR}_{\mathrm{standard}} = \frac{1}{M}\sqrt{\sum_{i=1}^{M}\left(\mathrm{SNR}_i - \mathrm{SNR}_{\mathrm{Average}}\right)^2} \tag{3.4}$$

$$\mathrm{SNR}_{\mathrm{Average}} = \frac{1}{M}\sum_{i=1}^{M}\mathrm{SNR}_i \tag{3.5}$$

where SNR_i is the SNR of the i-th signal.

A lower $\mathrm{SNR}_{\mathrm{standard}}$ value indicates greater consistency among the SNRs of signals originating from the same MS event. Figure 3.5c presents the distribution of $\mathrm{SNR}_{\mathrm{standard}}$ for the 74 rock fracturing events identified in this study (see Sect. 3.3.2).

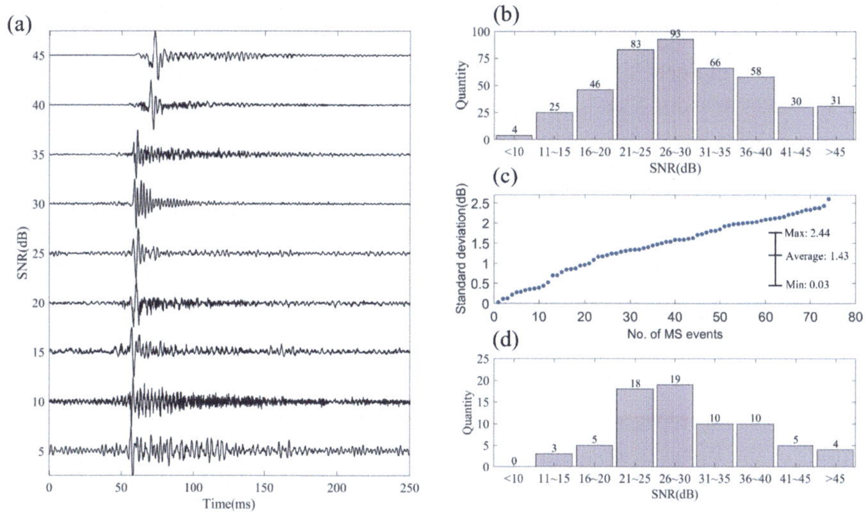

Fig. 3.5 The distribution characteristics of SNRs for MS signals and events recorded in the tunnel. It includes **a** representative MS signals with varying SNR levels, **b** the SNR distribution calculated from 436 MS signals, **c** standard deviations of SNRs for individual MS events, and **d** the distribution of average SNRs of MS events

The maximum observed $SNR_{standard}$ is only 2.4 dB, with an average of 1.4 dB, demonstrating that the SNR variation among signals within a given event is remarkably small. Consequently, it can be reasonably inferred that the SNR of any given MS signal is generally comparable to that of its corresponding SDDSR signal. Moreover, $SNR_{Average}$ serves as a reliable representative measure of the overall SNR for a specific MS event.

The distribution of $SNR_{Average}$ values is shown in Fig. 3.5d. Among the recorded MS events, only 29 (39.2%) exhibited SNRs exceeding 30 dB. Consequently, for the majority of events (60.8%), source locations derived without prior noise filtering may be unreliable due to inaccuracies in P-wave arrival picking. This finding underscores the critical motivation for the present study.

3.3 Methodology

3.3.1 Instantaneous Phase Difference

Previous research has demonstrated that MS events primarily take place in the proximity of the tunnel working face (Feng et al. 2013). Figure 3.6a depicts a schematic diagram of a typical tunnel configuration, in which a single MS event generates two signals that are recorded by sensors A and B positioned in different rows. Through the analysis, the following assumptions are posited: (i) the source waves propagate along rectilinear paths; (ii) the wave velocity within the rock mass is uniform; and (iii) the disparities in frequency attenuation between sensors B and A are deemed negligible, as elaborated in Sect. 3.3.2. Let l_A and l_B represent the propagation distances from the source to sensors A and B, respectively. Considering that $l_A > l_B$, the difference $l_A - l_B$ defines the source wave traveling path difference (SWTPD).

To illustrate the concept of the instantaneous phase, Fig. 3.6b presents idealized signals recorded by two sensors, each consisting of a single sinusoidal cycle. Consistent with the analysis in Sects. 3.3.2, the figure is configured such that the difference $l_A - l_B$ is clearly smaller than the wavelength L_p, i.e., SWTPD < L_P. Accordingly, the following relationship can be established:

$$t_{Ba} < t_{Aa} < t_{Be} < t_{Ae} \tag{3.6}$$

where t_{Aa} and t_{Ba} represent the arrival times. t_{Ae} and t_{Be} are the end times of the first periods of the P-waves at A and B, respectively.

The instantaneous phases $\omega_A(t)$ and $\omega_B(t)$ of the two MS signals are obtained by applying Hilbert transformations (HT) to the respective waveforms, yielding the results shown in Fig. 3.6c. The instantaneous phase difference $\Delta\omega(t)$ between the two signals is then derived as illustrated in Fig. 3.6d, expressed as:

$$\Delta\omega(t) = \omega_B(t) - \omega_A(t). \tag{3.7}$$

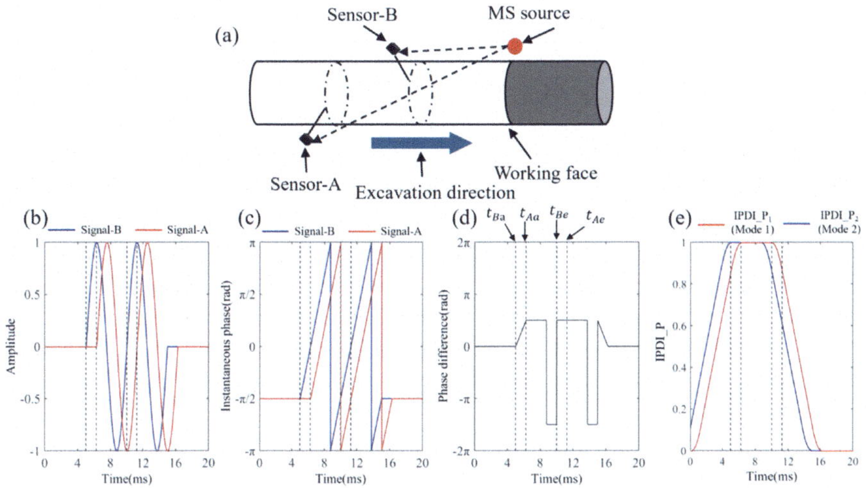

Fig. 3.6 Analysis of the signals recorded at sensors A and B. **a** Provides a schematic diagram illustrating the positions of the MS source and sensors. **b** Shows the idealized MS signals received at sensors A and B. The corresponding instantaneous phases of these signals are plotted in (**c**), while (**d**) displays their instantaneous phase difference. **e** Depicts the resulting IPDI_P$_1$ and IPDI_P$_2$ curves

As shown in Fig. 3.6d, $\Delta\omega(t)$ has one of two values in the interval $t_{\mathrm{Aa}} \le t \le t_{\mathrm{Be}}$. In this time window, either

$$\Delta\omega(t) = (l_{\mathrm{A}} - l_{\mathrm{B}})/L_{\mathrm{P}} \cdot 2\pi \tag{3.8}$$

or

$$\Delta\omega(t) = (1_{\mathrm{A}} - l_{\mathrm{B}})/L_{\mathrm{P}} \cdot 2\pi - 2\pi. \tag{3.9}$$

Let $\omega_{\mathrm{e}} = (d_{\mathrm{A}} - d_{\mathrm{B}})/L_{\mathrm{P}} \cdot 2\pi$. For a given MS event, the quantities l_{A}, l_{B}, and L_{P} are fixed, so ω_{e} takes a specific value accordingly. Based on the characteristics of the function $\Delta\omega(t)$, two forms of the function are constructed:

$$\text{Mode 1}: \quad \text{IPDI\_}P_1(t) = \int_t^{t+T} \left\{ |\Delta\omega(\tau) - \omega_{\mathrm{e}}|^2 + [2\pi - |\omega_{\mathrm{e}} - \Delta\omega|(\tau)]^2 \right\} \mathrm{d}\tau \tag{3.10}$$

$$\text{Mode 2}: \quad \text{IPDI\_}P_2(t) = \int_{t+\Delta t}^{t+\Delta t+T} \left\{ |\Delta\omega(\tau) - \omega_{\mathrm{e}}|^2 + [2\pi - |\omega_{\mathrm{e}} - \Delta\omega(\tau)|]^2 \right\} \mathrm{d}\tau \tag{3.11}$$

where

$$\Delta t = \frac{\omega_e}{2\pi} \times T \tag{3.12}$$

and T is the period of the P-wave and Δt is the time offset between the two signals.

Figure 3.6e illustrates the behavior of the IPDI_P curves over time for the considered case. Beginning at time t_{Ba}, the IPDI_P$_1$ curve increases steadily and attains its maximum value precisely at t_{Aa}. This distinct peak closely corresponds to the P-wave arrival time of the signal received later at sensor A (as depicted in Fig. 3.6b). Therefore, by evaluating the IPDI_P$_1$ function derived from the two MS signals, the P-wave arrival time at sensor A can be effectively identified. Similarly, the IPDI_P$_2$ curve enables the determination of the P-wave arrival time at sensor B, as also shown in Fig. 3.6e.

3.3.2 *The Process of IPDI_P*

Figure 3.7 illustrates the workflow of the IPDI_P method for identifying P-wave arrival times in tunnel MS monitoring. The procedure comprises four main steps, as outlined below.

Step 1 (signal marking): For a given MS event, the recorded signals are denoted as S_i ($i = 1, 2, \ldots, M$), with the index i reflecting the recording order. A smaller i indicates an earlier recording time and closer proximity of the corresponding sensor to the MS source.

Step 2 (suitability checking and auxiliary signal selection): The dominant frequency f_i of signal S_i is obtained via fast Fourier transform (FFT). Subsequently, the unified dominant frequency F_a and the P-wave wavelength L_P are computed using Eqs. 3.1 and 3.2. Based on the spatial configuration of the sensor array, the SDDSR associated with S_i is determined. If SDDSR $\leq L_P$, the corresponding SDDSR signal is selected as the auxiliary signal for P-wave arrival picking and is designated RS_i. Otherwise, the IPDI_P method is considered inapplicable to signal S_i.

Step 3 (IPDI_P calculation): Both S_i and RS_i are transformed via the Hilbert transform (HT) to extract their instantaneous phase information. Accurate estimation of the SWTPD between the two signals is challenging without prior arrival picking. Therefore, the spatial distance between the two sensors (i.e., SDDSR) is adopted as a surrogate. The ω_e is then estimated using the following expression:

$$\omega_e = \frac{\mathrm{SDDSR}}{L_P} \cdot 2\pi \tag{3.13}$$

The selection of the IPDI_P mode is determined by the chronological order of the signals S_i and RS_i. If S_i is recorded after RS_i, the IPDI_P$_1$ function defined in Eq. 3.10 is applied; otherwise, the IPDI_P$_2$ function is used. The parameter T is taken as $1/F_a$.

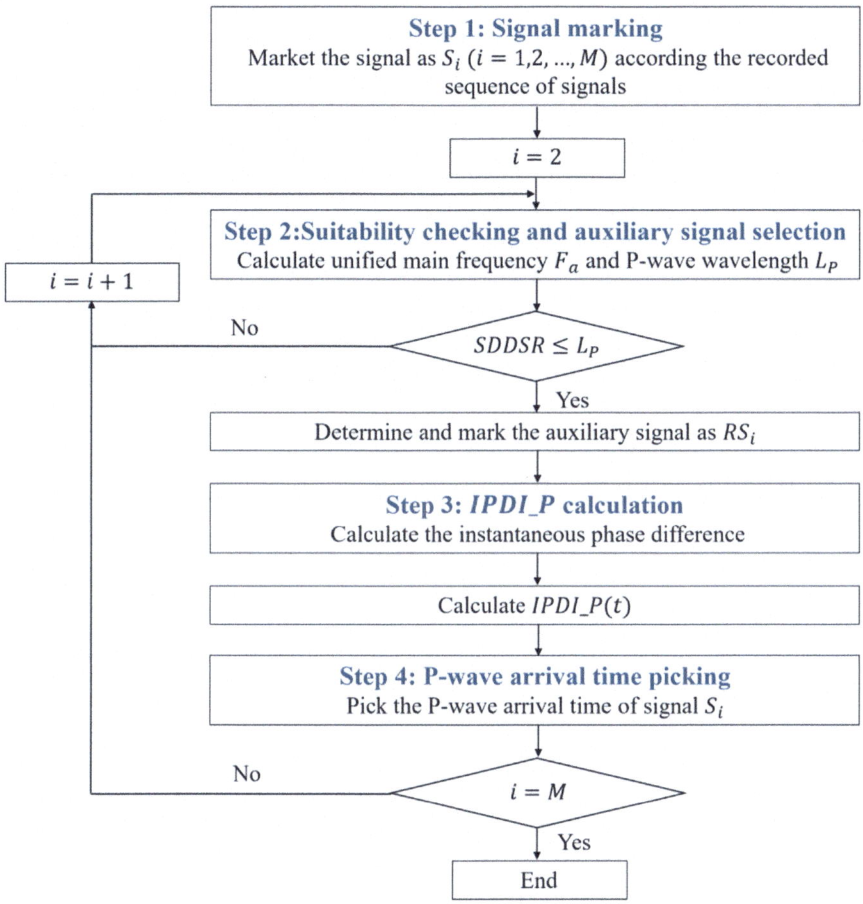

Fig. 3.7 The workflow of the IPDI_P method for identifying P-wave arrival times in tunnel MS monitoring

Step 4 (P-wave arrival time picking): In the IPDI_Pi curve, the time corresponding to the first peak, which is equal to or close to the maximum value, is selected as the P-wave arrival time of signal S_i.

3.4 Performance Analysis

3.4.1 Noise Simulation

This chapter evaluates the performance of the IPDI_P method through numerical simulations. To generate rock fracture signals with varying SNRs, it is first necessary to construct representative noise signals. Accordingly, the two primary noise sources (electrical and mechanical) discussed earlier are simulated.

The electrical noise signal EL(t) is synthesized by combining a deformed triangular wave TR(t) with a fundamental frequency of 50 Hz and a Gaussian white noise component GAW(t), as expressed in the following formulation:

$$\mathrm{EL}(t) = \beta_1 \cdot [\eta_1 \cdot \mathrm{TR}(t) + \eta_2 \cdot \mathrm{GAW}(t)] \tag{3.14}$$

where β_1 is a scaling factor, η_1 and η_2 are weighting coefficients, and

$$\mathrm{TR}(t) = \begin{cases} st - 1 & 0 \leq t < 2.5 \\ 0 & 2.5 \leq t < 7.5 \\ st - 3 & 7.5 \leq t < 10 \\ -st + 5 & 10 \leq t < 12.5 \\ 0 & 12.5 \leq t < 17.5 \\ -st + 7 & 17.5 \leq t \leq 20 \end{cases} \tag{3.15}$$

where s is the slope of an oblique line associated with the deformed triangular wave.

The mechanical vibration signal ME(t) is synthesized by summing four colored noise signals CO(t):

$$\mathrm{ME}(t) = \beta_2 \cdot \sum_{i=1}^{4} \mu_i \cdot \mathrm{CO}_i(t) \tag{3.16}$$

where β_2 is a scaling factor and μ_i denotes the weighting coefficient of the colored noise component $CO_i(t)$. The four types of colored noise are synthesized by filtering distinct random white noise sequences through bandpass filters corresponding to different frequency bands.

The parameters employed to generate the two types of noise (Table 3.1, Fig. 3.8) are specifically tailored to replicate the characteristics presented in Fig. 3.4. Using the simulated electrical noise signal as an illustration, its time-domain waveform (Fig. 3.8a) exhibits a periodic triangular pattern closely resembling the measured electrical noise signal (Fig. 3.4a). In the frequency domain, both the simulated and measured signals (Figs. 3.4b and 3.8b) share a dominant frequency of 50 Hz, with additional spectral components concentrated around similar frequencies, such as 250 Hz and 350 Hz. Furthermore, the amplitude distribution of the simulated

Table 3.1 The parameters for simulating the noise signals

Noise	Component	Simulation parameter		
		Characteristic	Value	Weighting coefficient
Electrical	TR	Slope of line, s :	0.4	1
	GAW	–		0.1
Mechanical	CO_1	Frequency range (Hz)	150–300	1.4
	CO_2	Frequency range (Hz)	0–600	1.12
	CO_3	Frequency range (Hz)	600–1200	0.22
	CO_4	Frequency range (Hz)	5–15	0.1

Both signals are scaled to a maximum amplitude of 1 m/s

noise (Fig. 3.8c) closely aligns with that of the measured signal (Fig. 3.4c) and the corresponding fitting parameters.

For a signal known to be composed of a pure rock-fracturing (PRF) signal and a noise (N) signal, the SNR can be calculated using the following formula:

$$\mathrm{SNR} = \mathrm{AP_{PRF}} - \mathrm{AP_N} \tag{3.17}$$

$$\mathrm{AP_{PRF}} = 10 \times \log_{10} P_{\mathrm{PRF}} \tag{3.18}$$

$$\mathrm{AP_N} = 10 \times \log_{10} \mathrm{P_N} \tag{3.19}$$

where $\mathrm{AP_{PRF}}$ ($\mathrm{AP_N}$) and $\mathrm{P_{PRF}}$ ($\mathrm{P_N}$) are the absolute power and power of the pure rock-fracturing (or noise) signal, respectively.

To generate an MS signal with a specified SNR, the PRF signal is first generated. Its absolute power is subsequently calculated using Eq. 3.18. The absolute power of the noise component is then determined according to Eq. 3.17, and the required noise power is derived via Eq. 3.19. By adjusting the parameters η_1 and η_2, a noise signal can be constructed to match the target power. Finally, the PRF signal and the synthesized noise signal are combined to produce a synthetic signal with the desired SNR.

3.4.2 Simulation Model and Parameters

A two-dimensional (2D) simulation model was created to assess the performance of the IPDI_P method together with two traditional methods, the STA/LTA and AIC methods (Fig. 3.9). In this model, an MS event is assumed to occur near the tunnel working face and is recorded by two sensors installed in different rows. The initial signals captured by the sensors are clean and noise-free. Subsequently, a range of MS signals with varying SNRs are generated following the procedure described in

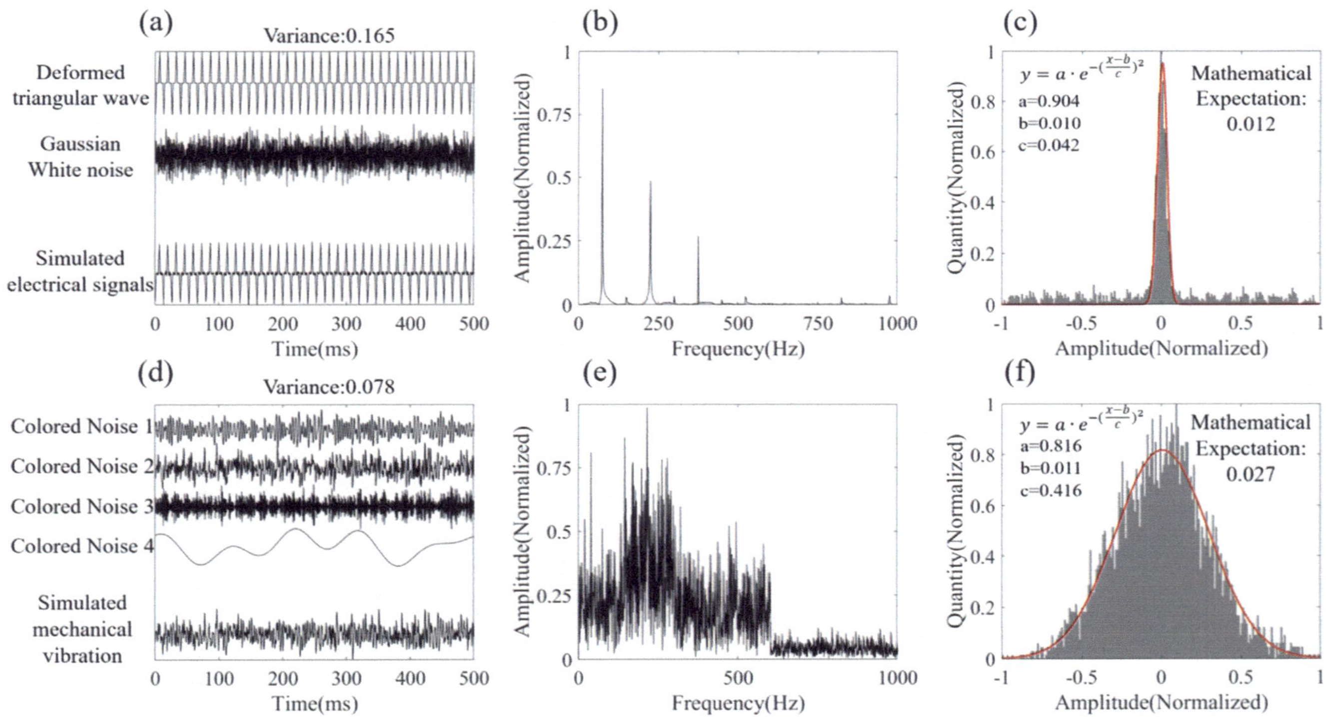

Fig. 3.8 The features of the simulated noise signals. **a–c** correspond to the simulated electrical noise signal, depicting its time-domain waveform, frequency-domain spectrum, and amplitude distribution, respectively. Similarly, **d–f** provide the corresponding representations for the simulated mechanical vibration signal

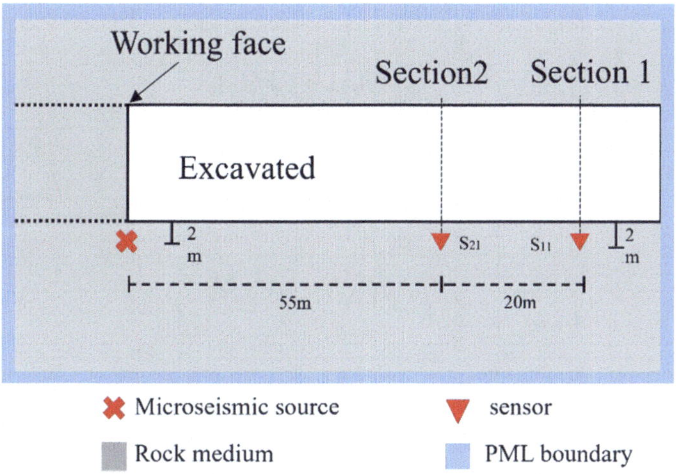

Fig. 3.9 The 2D model used in the numerical simulations

Table 3.2 Material parameters used in the simulations

Medium	Velocity (km/s)		Density (kg/m^3)	Attenuation coefficient (s^{-1})	
	P-wave	S-wave		P-wave	S-wave
Rock	6	3	2800	10	50
Tunnel (air)	0.34	0	7.9	5000	7000

Sect. 3.4.1. In all simulation cases, the SNRs of the signals recorded by the two sensors are configured to be identical.

The waves propagate based on the following 2D viscoelastic wave equations (with uniform velocity). These wave equations are resolved by employing the staggered-grid finite-difference method (Carcione and Helle 1999; Virieux 1986) along with the finite-difference time-domain method (Yee 1966). A perfectly matched layer (PML) absorbing boundary condition (Berenger 1994) is applied to facilitate the simulation of wave propagation in an unbounded domain within a finite computational region. The values of the other material parameters adopted are listed in Table 3.2.

We used the STA/LTA (Chen et al. 2019) and AIC (Maeda 1985) methods to pick the P-wave arrivals to compare with IPDI_P.

3.4.3 *Typical Simulation Case*

Using the model and parameters established in Sect. 3.4.2, two MS signals with an SNR of 20 dB are generated to illustrate the construction of noisy MS waveforms and

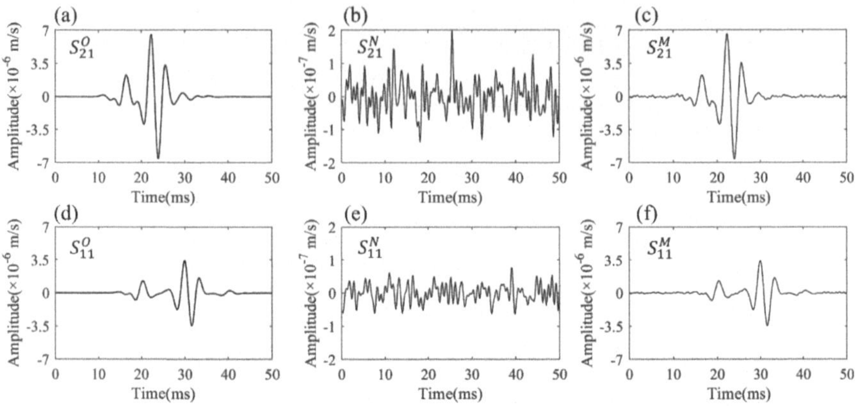

Fig. 3.10 Simulation of MS signals at an SNR of 20 dB. **a**, **d** Depict the pure signals S_{21}^{O} and S_{11}^{O}, respectively. The artificially generated mechanical vibration noise signals, with absolute power levels of − 137 and − 143 dBW, are shown in (**b**) and (**e**). By superimposing the corresponding noiseless signals and noise components, the mixed signals S_{21}^{M} and S_{11}^{M} are obtained, both exhibiting an SNR of 20 dB, as illustrated in (**c**) and (**f**)

to evaluate the comparative performance of the IPDI_P, STA/LTA, and AIC methods. The original noise-free signals, denoted as S_{21}^{O} and S_{11}^{O}, are shown in Figs. 3.10a, d, respectively. These waveforms correspond to the signals that would be recorded by the sensors in the absence of noise. Consequently, their P-wave arrival times are clearly distinguishable, occurring at 9.2 ms and 12.5 ms, as indicated.

The absolute powers of signals S_{21}^{O} and S_{11}^{O} are calculated as −117 and −123 dBW, respectively, using Eq. 3.18. To obtain noisy signals with an SNR of 20 dB, Eq. 3.17 indicates that noise signals with absolute powers of −137 and −143 dBW are needed. In this example, mechanical vibration is selected as the noise source. Based on Eq. 3.16 and the noise parameters provided in Table 3.1, the scale factor β_2 for the −137 dBW mechanical vibration noise is set to 8.4×10^{-7}, while that for the −143 dBW noise is set to 4.2×10^{-7}. The resulting artificial mechanical vibration signals, designated S_{21}^{N} and S_{11}^{N}, are shown in Fig. 3.10b, e, respectively. The final MS signals with an SNR of 20 dB, denoted S_{21}^{M} and S_{11}^{M}, are obtained by superimposing the noise signals onto the original clean signals $\left(S_{21}^{\mathrm{O}} + S_{21}^{\mathrm{N}}\right.$ and $\left.S_{11}^{\mathrm{O}} + S_{11}^{\mathrm{N}}\right)$, as illustrated in Fig. 3.10c, f.

Following the workflow illustrated in Fig. 3.7, the applicability of the IPDI_P method to the generated signals is first assessed. The FFT of signals S_{21}^{M} and S_{11}^{M} yields dominant frequencies of 224 and 210 Hz, respectively. Since the method requires a unified frequency for the signal pair, the average of these two values is adopted, resulting in a P-wave wavelength of 27.65 m. The shortest distance between the two sensor rows (SDDSR) is 20 m (Fig. 3.9), which is less than the calculated wavelength. Thus, the IPDI_P method is suitable for identifying the P-wave arrival times of both signals.

The instantaneous phases of S_{21}^{M} and S_{11}^{M} are then obtained via HT, and their phase difference is computed, as shown in Fig. 3.11a, f. For signal S_{11}^{M}, the IPDI_P$_1$ function is applied. The constant instantaneous phase difference α is determined as 4.54 rad, corresponding to $2\pi \times 20/27.65$ according to Eq. 3.13. Using this value, the IPDI_P1 curve for S_{11}^{M} is constructed (Fig. 3.11i). The P-wave arrival time is identified as the first peak approaching or reaching the maximum of this curve, yielding a value of 12.8 ms. For signal S_{21}^{M}, the IPDI_P$_2$ function is employed following the same procedure, resulting in an estimated arrival time of 9.5 ms (Fig. 3.11j).

For comparison, the STA/LTA and AIC methods were also applied directly to S_{21}^{M} and S_{11}^{M} (Fig. 3.11d, e, g, h). The estimated P-wave arrival times from all three methods are summarized in Table 3.3. The results indicate that the IPDI_P method provides estimates closer to the true arrival times, with identification errors 2.7 to 5 times smaller than those of the conventional methods. In this example, the IPDI_P method demonstrates superior accuracy.

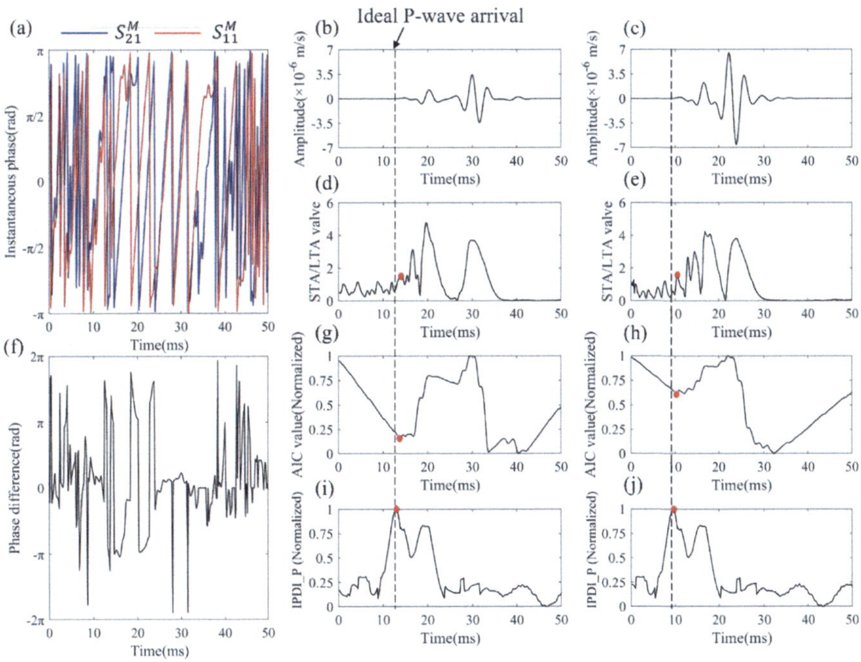

Fig. 3.11 The identification of P-wave arrival times for signals S_{21}^{M} and S_{11}^{M} is carried out using the IPDI_P, STA/LTA, and AIC methods. **a, f** present the instantaneous phases of the signals and their difference, respectively. The original noise-free signals are shown in (**b**) and (**c**). The analysis results for S_{11}^{M} using the STA/LTA, AIC, and IPDI_P methods are displayed in panels (**d**), (**g**), and (**i**), respectively. Correspondingly, panels (**e**), (**h**), and (**j**) illustrate the results obtained for S_{21}^{M} using the STA/LTA, AIC, and IPDI_P methods

Table 3.3 P-wave arrival times of the simulated MS signals (with SNRs of 20 dB) estimated using the IPDI_P, STA/LTA, and AIC methods

Simulated signal	True arrival time (ms)	Method	Calculated arrival time (ms)	
			Arrival time (ms)	Error (ms)
S_{21}^{M}	9.2	STA/LTA	10.5	1.3
		AIC	10	0.8
		IPDI_P	9.5	0.3
S_{11}^{M}	12.5	STA/LTA	14	1.5
		AIC	13.5	1
		IPDI_P	12.8	0.3

3.4.4 *Identifying P-Wave Arrival Times in More Complicated MS Signals*

To further evaluate the performance of the IPDI_P method, three sets of simulated signal pairs (S_{21}^{M} and S_{11}^{M}) were generated with SNRs ranging from 10 to 40 dB in 1 dB increments. The first set comprised 31 signal pairs contaminated solely by electrical noise, while the second set consisted of 31 pairs affected only by mechanical vibration noise. The third set was designed to represent more complex interference, containing a mixture of both electrical and mechanical noise. In this set, the absolute power difference between the two noise components was varied from –20 dB to 20 dB in steps of 1 dB. Consequently, a power difference of 20 dB corresponds to predominantly electrical noise, whereas –20 dB indicates mainly mechanical vibration. This configuration yielded a total of 1,271 signal pairs (41 power difference levels × 31 SNR levels). For each pair, the P-wave arrival times of S_{21}^{M} and S_{11}^{M} were estimated using the IPDI_P, STA/LTA, and AIC methods, all applied without any prior noise filtering.

Scenario 1: Single-Source Noise Interference

Figure 3.12 presents the picking errors obtained under single-noise-source conditions. As shown in Figs. 3.12a, c, when signals were contaminated by electrical noise, the errors for S_{21}^{M} and S_{11}^{M} were highly similar across methods. In general, picking errors increased as the SNR decreased. For signals with SNR $\geq$ 30 dB, all three methods maintained errors below 0.3 ms. However, as the SNR declined from 30 dB to 10 dB, the error associated with the STA/LTA method increased markedly, at a rate of approximately 0.3 ms per 5 dB. In contrast, the IPDI_P method exhibited a much slower error growth rate of approximately 0.1 ms per 5 dB, which is three times lower than that of STA/LTA. At an SNR of 10 dB, the error using IPDI_P was approximately 0.8 ms, roughly half that of STA/LTA. The AIC method consistently produced errors intermediate between those of the other two approaches.

For signals affected solely by mechanical vibration (Fig. 3.12b, d), the IPDI_P method again yielded the smallest errors at all SNR levels and exhibited the slowest error accumulation as the SNR decreased. When the SNR exceeded 30 dB, the errors

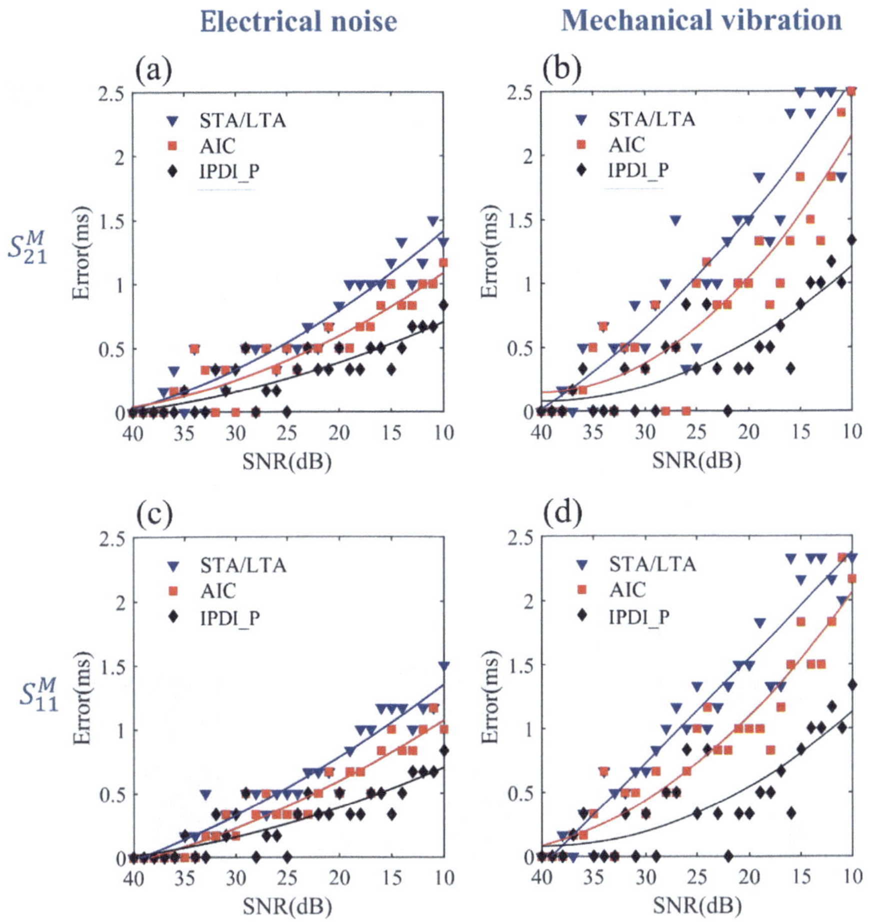

Fig. 3.12 The picking accuracy of P-wave arrival times obtained through the STA/LTA, AIC, and IPDI_P methods is evaluated and compared for microseismic signals contaminated by a single noise source. **a**, **c** Present the results for signals S_{21}^{M} and S_{11}^{M}, respectively, when both are affected solely by electrical noise. Correspondingly, panels (**b**) and (**d**) display the results for the same signals under mechanical vibration noise interference

from the IPDI_P method remained below 0.5 ms. However, as the SNR dropped, the error rates increased sharply for all methods. For instance, the STA/LTA method showed an error growth rate of 0.5 ms per 5 dB. At an SNR of 10 dB, both STA/LTA and AIC produced errors near 2.5 ms, whereas the IPDI_P method maintained a substantially lower error of approximately 1.3 ms.

It is also noteworthy that, for a given SNR, mechanical noise consistently induced larger picking errors than electrical noise, regardless of the method employed.

Scenario 2: MS Signals with Noise from Two Sources

Figure 3.13 shows representative outcomes from simulations incorporating mixed electrical and mechanical noise, with a specific focus on the P-wave arrival picking errors for signal S_{11}^{M} derived from the three methods. A typical waveform is provided in Fig. 3.13d, while the comprehensive simulation results are depicted as heatmaps in Fig. 3.13a–c, wherein a more extensive blue region signifies superior error mitigation. It is apparent that the IPDI_P method consistently achieves the lowest picking errors across all scenarios examined. Additionally, regardless of the chosen method, errors escalate with an increasing proportion of mechanical vibration noise (and a concomitant decrease in electrical noise).

In conclusion, the numerical simulations yield several important insights: (i) Lower SNRs are associated with greater errors in P-wave arrival identification. (ii) Among the three techniques assessed, the IPDI_P method exhibits the most robust overall performance in both accuracy and stability when applied to raw MS signals, particularly under SNR conditions below 30 dB. (iii) Mechanical vibration noise exerts a more pronounced adverse effect on picking precision compared to electrical noise.

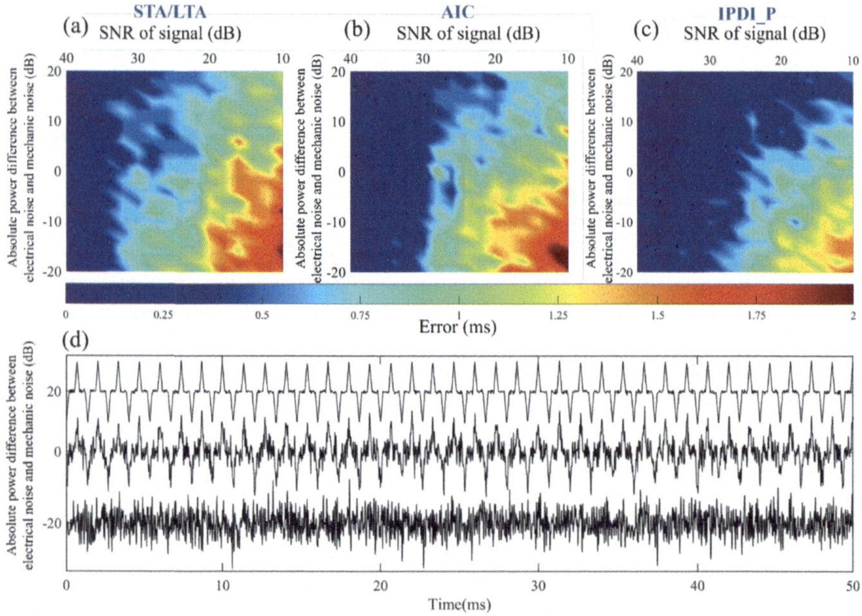

Fig. 3.13 The P-wave arrival picking errors for MS signals contaminated by noise sources. **a**–**c** Present the error distributions obtained with the STA/LTA, AIC, and IPDI_P methods, respectively. A representative noise waveform is displayed in panel (**d**), demonstrating the composite nature of the signal, which comprises both electrical and mechanical noise components at varying absolute power levels

3.5 Field Application

Application of the IPDI_P method to the Jinping tunnel data revealed that 400 of the 436 recorded rock fracture MS signals were suitable for further analysis, corresponding to 68 distinct MS events. The P-wave arrival times of these signals were subsequently estimated using the IPDI_P, AIC, and STA/LTA methods. These arrival picks were then input into seismogram processing software (supplied by the MS equipment manufacturer, Integrated Seismic System) to locate the 68 events; the resulting locations are presented in Fig. 3.14.

The MS events were categorized into two groups based on $SNR_{Average}$. Type-1 events, comprising 29 events, exhibited $SNR_{Average}$ values exceeding 30 dB. The remaining 39 events, which did not meet this threshold, were designated type-2 events. Figure 3.14a–c show the locations of type-1 events determined using the three picking methods, with each sphere representing an individual event. The locations derived from all three methods are highly consistent. Notably, the location of the 7/21 rockburst event (characterized by an SNR of 51 dB) was identical regardless of the method employed.

Previous research has established that MS events tend to cluster in the vicinity of an impending strong rockburst prior to its occurrence (Feng et al. 2013; He et al. 2019; Cai et al. 2020). Accordingly, the spatial distributions of the located events relative to the 7/21 rockburst were analyzed, with the results summarized in Table 3.4. For type-1 events, the average distance from the event centroids to the rockburst location ranged from 5.18 to 5.71 m, depending on the picking method used. These values are closely comparable, and the relative accuracy among methods aligns with findings from prior studies.

However, type-2 events constituted the majority (57%) of the 68 events analyzed. Figure 3.14d–f illustrate the locations of these events as determined by the STA/LTA, AIC, and IPDI_P methods, respectively. Given that both event types are temporally associated with the 7/21 rockburst, the centroid of type-2 events would be expected to lie at a distance from the rockburst comparable to that of type-1 events. As shown in Table 3.4, the STA/LTA and AIC methods yielded centroid distances for type-2 events that were at least 2.5 times greater than those for type-1 events, indicating unreliable P-wave arrival picks for these lower-SNR signals. In contrast, the IPDI_P method located the majority of type-2 events in close proximity to the 7/21 rockburst, with a centroid distance of 5.61 m, which is similar to that achieved for type-1 events. This result demonstrates that the P-wave arrival times identified by the IPDI_P method remain reliable even for events with lower SNRs.

This field case demonstrates that the IPDI_P method provides reliable estimates of MS event locations and effectively captures the incubation process of rockbursts in tunnels. Furthermore, for events with SNRs below 30 dB, the IPDI_P method exhibits greater reliability than the STA/LTA and AIC methods.

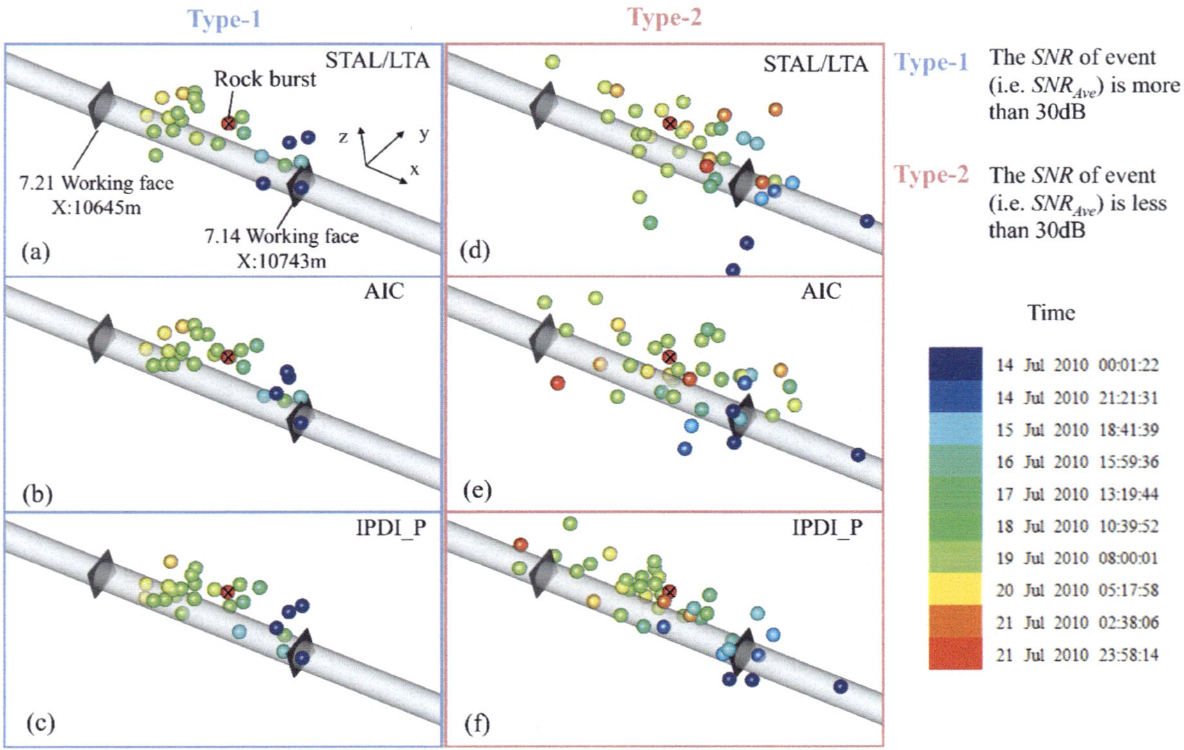

Fig. 3.14 The spatial distribution of MS events occurring in the Jinping tunnel, as located by the three distinct methods. **a**–**c** depict the locations of type-1 MS events determined using the STA/LTA, AIC, and IPDI_P methods, respectively. Similarly, (**d**)–(**f**) show the corresponding location results for type-2 MS events obtained with the same three methods in the same order

Table 3.4 Assessment of the degree of aggregation of the MS events associated with the 7/21 rockburst (located at $x = 10703.93$ m, $y = 8.07$ m, $z = -29.13$ m)

Class	SNR_{Ave} (dB)	Method used	Center of MS events (m)			Distance of rockburst from center of the MS events (m)
			x	y	z	
Type-1	> 30	STA/LTA	10702.14	3.06	- 29.95	5.38
		AIC	10699.73	5.06	- 29.45	5.18
		IPDI_P	10700.66	3.46	- 29.96	5.71
Type-2	≤ 30	STA/LTA	10720.05	–1.93	- 33.87	19.55
		AIC	10716.92	5.52	- 31.21	13.40
		IPDI_P	10702.29	4.44	- 33.08	5.61

3.6 Discussion

The IPDI_P method has two fundamental prerequisites: the availability of an auxiliary signal corresponding to the target signal and a sensor spacing smaller than the P-wave wavelength. This approach is particularly well suited for tunnel MS monitoring, as the typical sensor array configuration in tunnels inherently satisfies these conditions (Fig. 3.1). In contrast, applying the method to large-scale underground excavations such as mines or cavern complexes (Ghosh and Sivakumar 2018) is often challenging because intersensor distances frequently exceed the P-wave wavelength, making it difficult to identify a suitable auxiliary signal. Therefore, the applicability of the IPDI_P method must be assessed based on site-specific conditions. When deployment is feasible, arranging sensors in closely spaced pairs is recommended to facilitate their use.

A critical step in implementing the IPDI_P method is the selection of the auxiliary signal. For signals recorded by sensors near the MS source, this selection is straightforward. However, for sensors farther from the source, multiple candidates exist. For instance, the final signal in a sequence of N recordings could theoretically use any of the preceding N–1 signals as its auxiliary. To investigate this issue systematically, two representative selection strategies were compared using data from the Jinping tunnel (Sect. 3.5): (i) selecting the auxiliary signal from the nearest sensor in the same row as the target and (ii) selecting it from the nearest sensor in a different row.

The analysis focused on type-1 MS events ($SNR_{Ave} > 30$ dB), whose clear waveforms permit accurate manual P-wave picking. A total of 97 target signals satisfying the IPDI_P criteria were examined. As shown in Fig. 3.15, option (i) yielded an average picking error of 1.2 ms, whereas option (ii) reduced the average error to 0.3 ms. These results demonstrate that selecting the auxiliary signal from a sensor in a different row significantly improves accuracy. Consequently, this strategy was adopted in the methodological framework presented in Sect. 3.3.

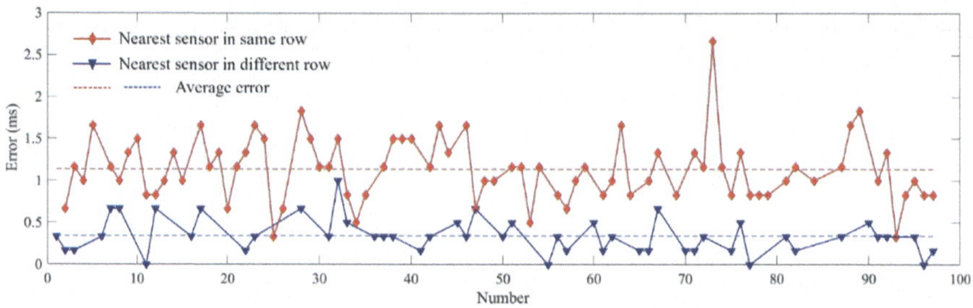

Fig. 3.15 Errors incurred using different auxiliary signals in the IPDI_P calculation

3.7 Conclusions

This study aims to enhance the accuracy of directly identifying P-wave arrival times in rock-fracturing MS signals recorded in deep buried tunnels, particularly under SNR conditions. The SNR evaluated over the full signal length tends to overestimate the P-wave SNR; in practice, over half of tunnel MS signals exhibit SNRs below 30 dB, corresponding to P-wave SNRs in the range of 16–24 dB. To address this challenge, the IPDI_P method, which leverages instantaneous phase information, was developed.

The method is founded on two key analytical insights: the source wave traveling path difference (SWTPD) between the target signal and its auxiliary signal and the shortest distance between different sensor rows (SDDSR). For most MS signals, the SDDSR is smaller than the P-wave wavelength, and the instantaneous phase difference between the P-waves of the target and auxiliary signals remains stable over specific time intervals. Two primary noise sources, electrical interference and mechanical vibration, were considered. Their time- and frequency-domain characteristics were analyzed and used to generate realistic noise samples for numerical simulations. Synthetic MS signals with SNRs ranging from 10 to 40 dB were subsequently produced and processed using the IPDI_P, STA/LTA, and AIC methods, all applied without prior noise filtering.

The results demonstrate that the IPDI_P method consistently outperforms the traditional STA/LTA and AIC methods in both stability and accuracy. Identification errors generally increase with decreasing SNR, with mechanical vibration noise inducing larger errors than electrical noise. The method was further validated using field data from the Jinping II hydropower station tunnel, where it provided more coherent localization of MS events preceding the 7/21 rockburst. Both numerical and field results confirm that IPDI_P offers superior reliability and precision for signals with SNRs below 30 dB.

By yielding more accurate P-wave arrival estimates, the IPDI_P method significantly improves the location of rockburst-related MS events. This enhanced spatial accuracy contributes to a better understanding of rockburst nucleation processes and supports the development of more reliable MS-based early warning systems. A notable strength of the method is its suitability for tunnel environments; however, this specificity may also limit its broader applicability. Further research is needed to assess the adaptability of the IPDI_P method to other engineering contexts, such as mining and slope excavation.

References

Akaike H (1974) A new look at the statistical model identification. IEEE Trans Automat Contr8 19(6):716–723

Akram J, Peter D, Eaton D (2018) A k-mean characteristic function to improve STA/LTA detection. In: Proceedings of the GeoConvention, pp 7–11

Allen RV (1978) Automatic earthquake recognition and timing from single traces. Bull Seismol Soc Am 68(5):1521–1532

Berenger JP (1994) A perfectly matched layer for the absorption of electromagnetic waves. J Comput Phys 114(2):185–200

Cai W, Bai X, Si G, Cao W, Gong S, Dou L (2020) A monitoring investigation into rock burst mechanism based on the coupled theory of static and dynamic stresses. Rock Mech Rock Eng 53:5451–5471

Carcione JM, Helle HB (1999) Numerical solution of the poroviscoelastic wave equation on a staggered mesh. J Comput Phys 154(2):520–527

Chen BR, Wu H, Chi XW, Liu H, Wu MD, Yan JW (2019) Real-time recognition algorithm for microseismic signals of rock failure based on STA/LTA and its engineering application. Rock Soil Mech 40(9):3689–3696

Dhanuka R, Singh JP, Tripathi A (2023) A comprehensive survey of deep learning techniques in protein function prediction. IEEE/ACM IEEE ACM Trans Comput Biol

Eisner L, Fischer T, Rutledge JT (2009) Determination of S-wave slowness from a linear array of borehole receivers. Geophys J Int 176(1):31–39

Fard RA, Javan Doloei G, Farrokhi M, Rahimi H, Mahood M (2020) Coda wave attenuation's dependency on Lapse time and frequency in west of Iran plateau using local earthquakes. Ann Geophys-Italy 63(4):437–437

Feng Q, Han L, Zhao B, Li Q (2023) Microseismic events recognition via joint deep clustering with residual shrinkage dense network. IEEE T Geosci Remote

Feng XT, Chen BR, Zhang CQ, Li SJ, Wu SY (2013) Mechanism, warning and dynamic control of rockburst development process. Science Press, Beijing

Ghosh GK, Sivakumar C (2018) Application of underground microseismic monitoring for ground failure and secure longwall coal mining operation: a case study in an Indian mine. J Appl Geophys 150:21–39

Gong J, Li Y, Wu N, Li M (2019) Automatic time picking of microseismic data based on shearlet-AIC algorithm. J Seismol 23:261–269

He S, Song D, Li Z, He X, Chen J, Li D, Tian X (2019) Precursor of spatiotemporal evolution law of MS and AE activities for rock burst warning in steeply inclined and extremely thick coal seams under caving mining conditions. Rock Mech Rock Eng 52:2415–2435

He Z, Peng P, Wang L, Jiang Y (2020) PickCapsNet: capsule network for automatic P-wave arrival picking. IEEE Geosci Remote S 18(4):617–621

Khalqillah A, Isa M, Muksin U (2018) A GUI based automatic detection of seismic P-wave arrivals by using short term average/long term average (STA/LTA) method. J Phys: Conf Ser 1116(3):032014

Li X, Shang X, Morales Esteban A, Wang Z (2017) Identifying P phase arrival of weak events: The Akaike information criterion picking application based on the empirical mode decomposition. Comput Geosci 100:57–66

Li X, Shang X, Wang Z, Dong L, Weng L (2016) Identifying P-phase arrivals with noise: an improved Kurtosis method based on DWT and STA/LTA. J Appl Geophys 133:50–61

Liu JP, Si YT, Zhang CY, Wang R (2020) Estimation of fracturing mechanisms by the ratio of radiated energy between S and P waves of microseismic events during mining in metal mines. Appl Geophys 17(3):465–474

Liu L, Li S, Xiao Y et al (2023) An inverted heterogeneous velocity model for microseismic source location in deep buried tunnels. Rock Mech Rock Eng 56(7):4855–4880

Maeda N (1985) A method for reading and checking phase times in autoprocessing system of seismic wave data. Zisin 38:365–379

Morita Y (1984) Automatic detection of onset time of seismic waves and its confidence interval using the autoregressive model fitting. Earthquake 37:281–293

Nejad FP, Jaksa MB (2017) Load-settlement behavior modeling of single piles using artificial neural networks and CPT data. Comput Geotech 89:9–21

Sabbione JI, Velis DR (2013) A robust method for microseismic event detection based on automatic phase pickers. J Appl Geophys 99:42–50

Song F, Toksöz MN (2011) Full-waveform based complete moment tensor inversion and source parameter estimation from downhole microseismic data for hydrofracture monitoring. Geophysics 76(6):103–116

Vaezi Y, Van der Baan M (2015) Comparison of the STA/LTA and power spectral density methods for microseismic event detection. Geophys Suppl Monthly No Roy Astronom Soc 203(3):1896–1908

Virieux J (1986) P-SV wave propagation in heterogeneous media: Velocity-stress finite-difference method. Geophysics 51(4):889–901

Withers M, Aster R, Young C, Beiriger J, Harris M, Moore S, Trujillo J (1998) A comparison of select trigger algorithms for automated global seismic phase and event detection. B Seismol Soc Am 88(1):95–106

Wong J, Han L, Bancroft J et al (2009) Automatic time-picking of first arrivals on noisy microseismic data. CSEG 1(1.2):1–4

Xiao Y, Chen S, Feng G et al (2025) Picking P-wave arrival times from microseismic signals in tunnels with low signal-to-noise ratios without noise filtering. Soil Dyn Earthquake Eng 192:109288

Xiao Y, Chen S, Liu L et al (2024) Determining S-wave arrival times for microseismic monitoring in tunnels based on instantaneous phase difference of sensors. Rock Mech Rock Eng 57(3):1781–1797

Xiao Y, Chen S, Liu L et al (2026) A stratified double-difference dynamic tomography of 3d velocity field for deep-buried large underground cavern: Y. Xiao et al. Rock Mech Rock Eng:1–24

Xiao YX, Feng XT, Chen BR et al (2015) Evolution of frequency spectrum during instant rockbursts in deep inoculation tunnel. Rock Soil Mech 36(4):1127–1134

Xiao YX, Feng XT, Hudson J et al (2016) ISRM suggested method for in situ microseismic monitoring of the fracturing process in rock masses. Rock Mech Rock Eng 49:343–369

Xiao Y, Chen S, Liu L, Li S, Chen B, Feng G (2026) Location for microseismic event clusters in deep tunnels based on coda wave interference. Tunnell Underground Space Technol 174:107690

Xu NW, Li TB, Dai F et al (2016) Microseismic monitoring of strainburst activities in deep tunnels at the Jinping II Hydropower Station, China. Rock Mech Rock Eng 49:981–1000

Yee K (1966) Numerical solution of initial boundary value problems involving Maxwell's equations in isotropic media. IEEE Trans Antennas Propag 14(3):302–307

Zhang W, Feng XT, Bi X et al (2021) An arrival time picker for microseismic rock fracturing waveforms and its quality control for automatic localization in tunnels. Comput Geotech 135:104175

Zheng J, Lu J, Peng S et al (2018) An automatic microseismic or acoustic emission arrival identification scheme with deep recurrent neural networks. Geophys J Int 212(2):1389–1397

Chapter 4
Determining S-Wave Arrival Times for Near-Field MS Data in Tunnel

4.1 Prologue

Accurate location of MS sources is essential for tracking rock fracture processes and identifying disaster precursors (Feng et al. 2019a, b; Xiao et al. 2024, 2025). The precision of MS localization depends primarily on two factors: the accuracy of arrival time picking for MS waves and the appropriateness of the velocity model for the monitored region (Zhu et al. 2013; Li et al. 2014; Liu et al. 2023; Wang et al. 2023). MS waveforms typically comprise P-waves, S-waves, and converted phases (Liu et al. 2019). P-wave arrival times are generally more straightforward to identify, as P-waves travel faster and thus arrive earliest following an MS event (Frid and Kamai 2020). In contrast, S-wave arrival identification is often hindered by interference from preceding P-waves and other waveforms, a challenge frequently encountered in near-field MS monitoring (Kuperkoch et al. 2012; Feng et al. 2017). Accurate S-wave picking is critical for improving event location accuracy and enhancing the robustness of velocity model-based localization algorithms (Kennett and Engdahl 1991; Wang et al. 2022). Addressing this challenge is therefore imperative for advancing disaster early warning capabilities.

The methods for S-wave arrival picking are largely adapted from those used for P-wave arrival picking, as described in Sect. 3.1. However, the extraction of S-wave arrivals is more challenging, particularly in the context of near-field MS monitoring. The overlap between the onset of the S-wave and the coda of the P-wave complicates the accurate identification of the arrival time. This chapter continues to employ the method proposed in Chap. 3 for S-wave arrival picking (instantaneous phase difference intensity of S-waves, IPDI_S). The method is first validated using analytical signals and numerically simulated data representative of tunnel MS monitoring conditions. Subsequently, it is applied to locate MS sources and rockburst events in the Jinping tunnel.

© Tongji University Press Co., Ltd 2026

Y. Xiao et al., *Near-Field Microseismic Data Processing in Deep Tunnels*,
https://doi.org/10.1007/978-981-92-3208-6_4

4.2 The Process of IPDI_S

Figure 4.1 illustrates a typical configuration of MS sensors in tunnel engineering projects. The configuration and movement of the sensors are identical to those described in Chap. 3. However, to facilitate the explanation of the IPDI_S, a different labeling scheme is adopted. The sensor labeling convention reflects this spatial arrangement. Each sensor is designated S_{ij} ($i = 1,2\ldots, p; j=1,2\ldots, q$), where i denotes the section number along the tunnel axis and j indicates the sensor number within that section.

Taking into account the layout and working conditions in the MS monitoring process (in tunnels), the IPDI_S method is carried out in four steps, as shown in Fig. 4.2 and described below.

(1) *Sequencing of the signals*: When the M sensors are triggered, the signals are arranged based on the time sequence in which they are activated by the MS event. The MS signals captured by the sensors are then labeled in the format Q_i ($i = 1, 2, \ldots, M$), where Q_i stands for the signal captured by the initially triggered sensor and Q_M stands for the signal captured by the finally triggered sensor. The trigger time is the P-wave arrival time, which can be selected manually.

(2) *Feasibility checking*: To ascertain the S-wave arrival time of the signal Q_i , the period T_i and wavelength L_i of the S-wave are calculated first. The formula is shown in Sect. 3.3. The P-wave velocity needs to be replaced with the S-wave velocity. The expression is then used to compute the propagation distance difference (P) between signal Q_i and signal Q_{i-1}.

$$P = (t_i - t_{i-1}) \times V_{\mathrm{P}} \tag{4.1}$$

where t_i and t_{i-1} denote the trigger times of the sensors associated with signals Q_i and Q_{i-1}, respectively, and V_{P} represents the P-wave velocity. The S-wave arrival time of signal Q_i can be selected via the IPDI_S method when $P < L_i$.

(3) *Calculate the IPDI_S*: The signals S_i and S_{i-1} are decomposed via ensemble empirical mode decomposition (EEMD) and rebuilt as $Q_i^{'}$ and $Q_{i-1}^{'}$ (Wu et al. 2009). The aimed—for phase difference ω_{e} between them is then computed using the formula:

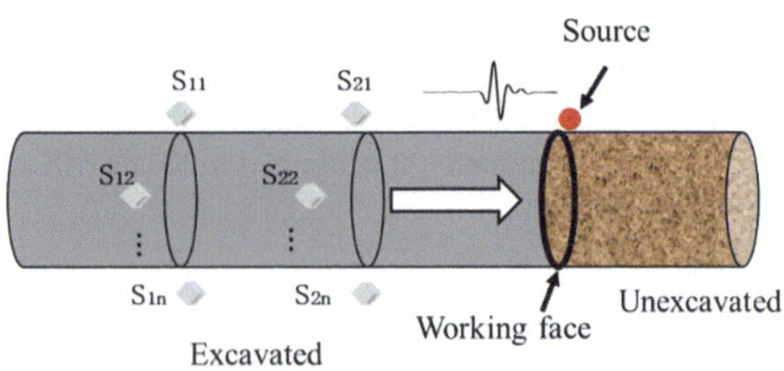

Fig. 4.1 Schematic diagram illustrating MS monitoring in tunnels

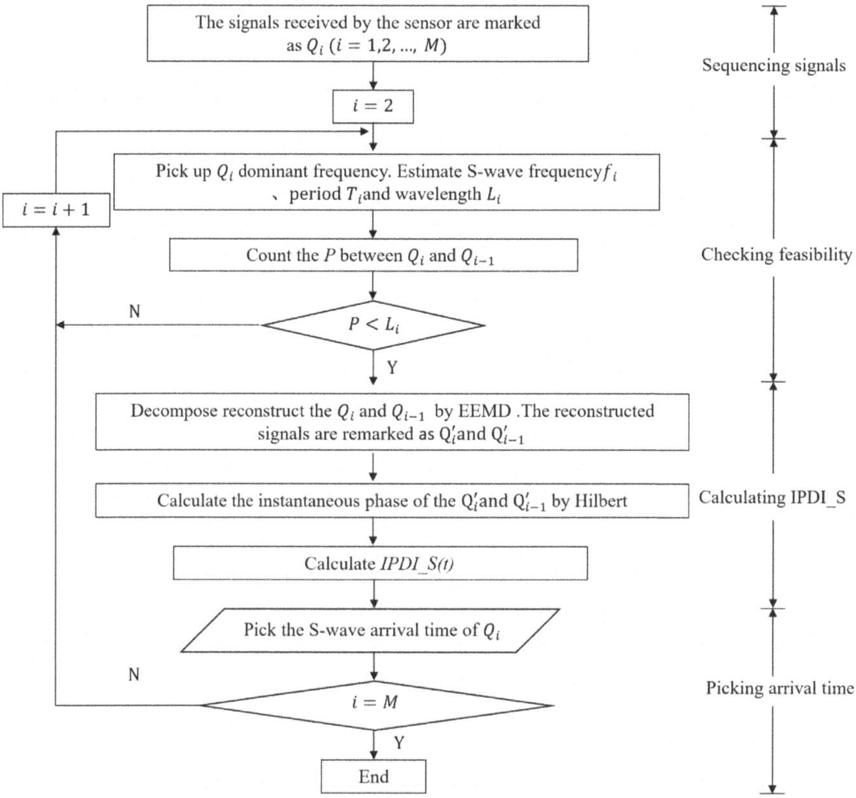

Fig. 4.2 Workflow used to implement the IPDI method

$$\omega_e = \frac{P}{L_i} \times 2\pi. \tag{4.2}$$

The instantaneous phases $\omega_i(t)$ and $\omega_{i-1}(t)$ of the reconstructed signals Q_i' and Q_{i-1}' are then computed, and the instantaneous phase difference $\Delta\omega(t)$ is calculated:

$$\Delta\omega(t) = \omega_{i-1}(t) - \omega_i(t). \tag{4.3}$$

IPDI_S can then be obtained using Eqs. 4.4–4.6.

$$\text{IPDI\_}S(t) = \int_t^{t+T} \left(\varphi_1{}^2 + \varphi_2{}^2\right)\mathrm{d}t \tag{4.4}$$

$$\varphi_1 = |\Delta\omega(t) - \omega_e| - 0 \tag{4.5}$$

$$\varphi_2 = 2\pi - |\Delta\alpha(t) - \omega_e|. \tag{4.6}$$

(4) *Picking the arrival time*: We find the maximum value point in IPDI_S and pick the corresponding time to be the S-wave arrival time of signal Q_i .

4.3 Numerical Validation

4.3.1 Numerical Simulation Method

2D numerical simulations are performed to validate the feasibility of the IPDI_S method. In this setup, six sensors arranged in two groups and five MS source locations are configured, as illustrated in Fig. 4.3. To model the propagation of MS waves, viscoelastic wave equations are formulated with appropriate damping coefficients, enabling the simulation of both longitudinal and transverse wave attenuation (Carcione 1999).

In this study, viscoelastic wave equations are solved by integrating the staggered-grid finite-difference method with the finite-difference time-domain technique. The solver strategy is the same as that in Sect. 3.4. The source time function is represented by a Ricker wavelet. The velocities of the surrounding rock are 5000 m/s (P-wave) and 3000 m/s (S-wave).

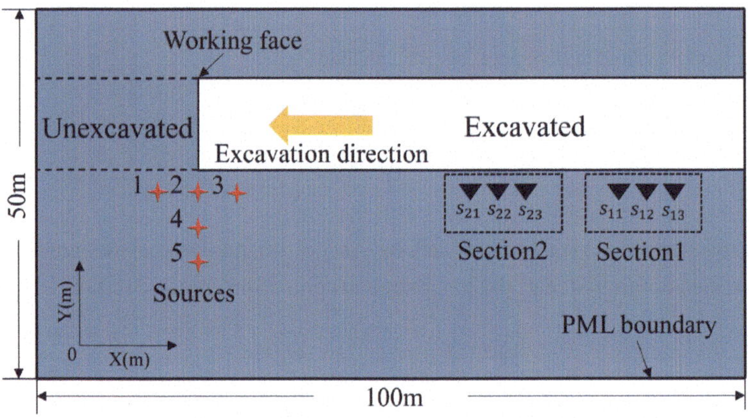

Fig. 4.3 2D numerical simulation of an MS monitoring experiment in a tunnel

4.3.2 S-Wave Arrival Time Picking Based on the IPDI Method

Figure 4.4 presents the waveforms recorded by the six sensors during the numerical simulation using the third ms source. Given that the simulation incorporates no noise and employs an ideal Ricker wavelet as the source, the arrival times of both P- and S-waves are relatively distinguishable overall. Nevertheless, even under such idealized conditions, a substantial degree of overlap between the P- and S-wave phases is observable at the sensor locations, complicating the precise identification of the S-wave onset.

In the following analysis, the signal captured by sensor S_{23} is used as an example to demonstrate the application of the IPDI method for determining the S-wave arrival time.

First, the signals are arranged in chronological order according to sensor trigger times. For the third MS source, the sequences are S_{21}, S_{22}, S_{23}, S_{11}, S_{12}, and S_{13}. To determine the S-wave arrival time at sensor S_{23}, the signals from sensors S_{22} and S_{23} are selected for analysis (Fig. 4.5a).

The applicability of the IPDI_S method for S-wave picking is assessed. The FFT of the signal recorded by S_{23} yields a dominant frequency of 304 Hz (Fig. 4.5b). Given an S-wave velocity of 3000 m/s, the corresponding S-wave wavelength is calculated as 9.87 m (3000/304), and the period is 3.29 ms (1/304). Using the P-wave arrival time and velocity (5000 m/s), P is determined to be 2.5 m. Since P is smaller than the S-wave wavelength, the IPDI_S method is suitable for identifying the S-wave arrival.

The IPDI_$S(t)$ function is then computed. Owing to the absence of noise in the simulation, the signal reconstructed via EEMD closely matches the original waveform (Fig. 4.5c). These reconstructed signals are calculated by HT to derive their

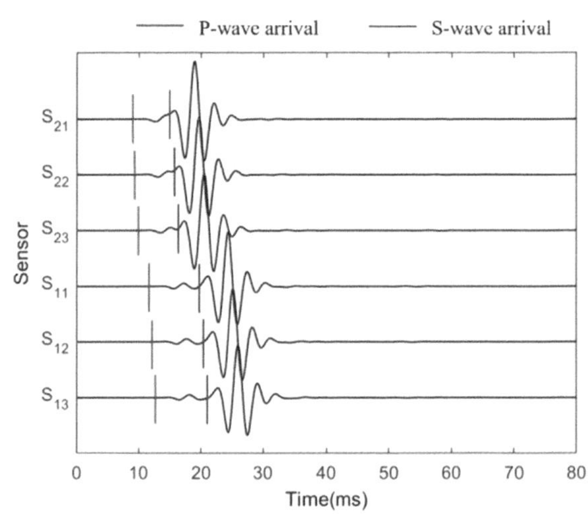

Fig. 4.4 Simulated signals recorded from an MS event at source 3

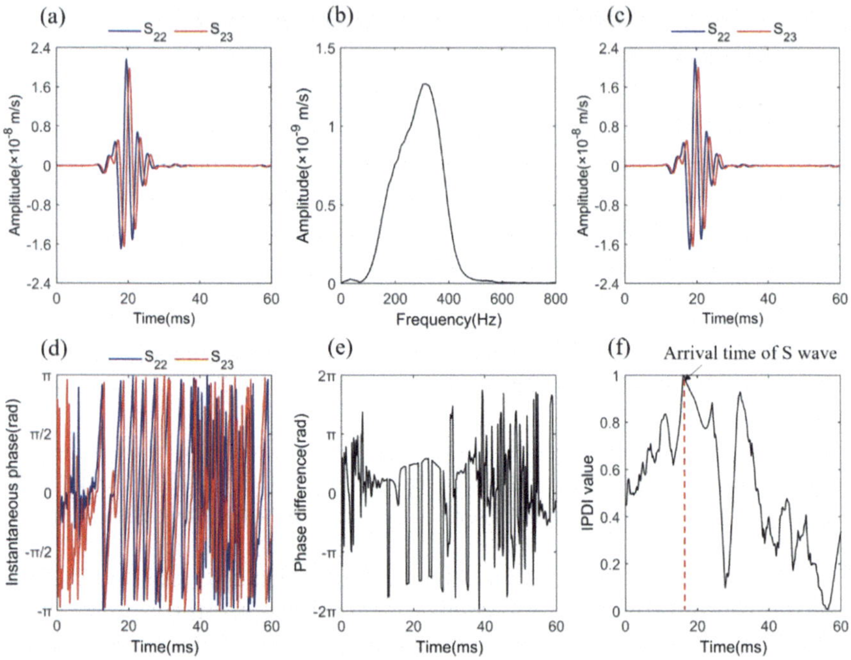

Fig. 4.5 Analysis of the signal received by sensor S_{23} by IPDI. **a** The signals captured by the two sensors. **b** The frequency spectrum associated with S_{23}. **c** Reconstructed signals for the two sensors acquired through EEMD. **d** Instantaneous phases of the two reconstructed signals. **e** Instantaneous phase difference between the two reconstructed signals. **f** IPDI_S curve

instantaneous phases (Fig. 4.5d) and the phase difference (Fig. 4.5e). Incorporating the calculated P into Eq. 4.2 gives $\omega_e = 73°$. This value is then used in Eqs. 4.4–4.6 to generate the IPDI_$S(t)$ curve (Fig. 4.5f). The time corresponding to the maximum of this curve is identified as 16.3 ms.

Given that the coordinates of both sources and sensors are precisely known in the numerical simulation, the theoretical S-wave arrival time can be computed directly from the propagation distance. For the signal received at S_{23} from the third source, this theoretical arrival time is 16.3 ms, which exactly matches the value obtained using the IPDI_S method.

For each of the five MS sources positioned at distinct locations, every sensor records five signals, enabling the IPDI_S method to be applied for S-wave arrival picking in subsequent analysis. Since the method requires paired sensors, a total of 25 S-wave arrival times are obtained using the following pairings: S_{22} with S_{21}, S_{23} with S_{22}, S_{11} with S_{23}, S_{12} with S_{11}, and S_{13} with S_{12}. Fig. 4.6 displays the maximum, average, and minimum picking errors for each sensor. The results indicate that the maximum error across all picks is 1 ms, the minimum error is 0 ms, and the overall average error for the 25 picks is 0.44 ms. These findings demonstrate that the IPDI_S

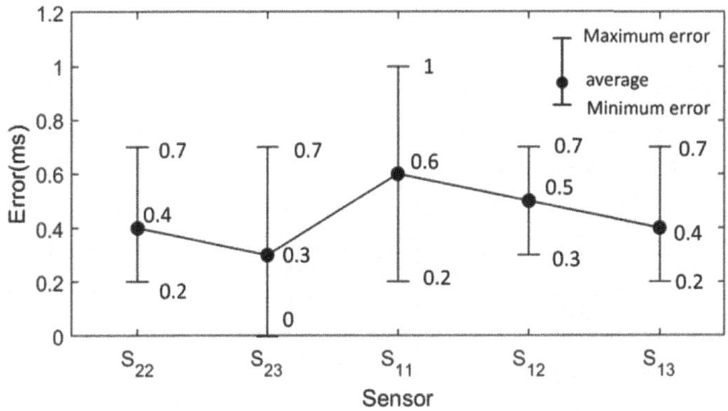

Fig. 4.6 Absolute errors found when using the IPDI method to select S-wave arrival times

method yields consistently small and stable errors, confirming its viability for S-wave arrival picking in tunnel environments.

4.4 Field Study

4.4.1 Background

The Jinping II hydropower station is located on the main course of the Yalong River in Sichuan Province, China. The project comprises seven parallel tunnels with an average burial depth of approximately 1610 m and a maximum depth of approximately 2525 m. Stress measurements indicate that once the burial depth exceeds 1200 m, the stress field transitions from a locally stressed condition within the valley slope zone to a self-weight stress field dominated by vertical principal stresses. The stress magnitudes increase with depth, and the maximum principal stress in the engineering area reaches 63 MPa, classifying it as a high-stress zone. Consequently, the likelihood of rockburst occurrence in these tunnels is significantly elevated. Figure 4.7 illustrates three tunnel sections that experienced successive severe rockbursts, highlighting the substantial threat to worker safety (Niu et al. 2021). Thus, implementing MS monitoring is essential for assessing tunnel conditions and providing early warning of potential rockburst events.

The MS monitoring system employed at the station consists of three main components: an external control center, internal data acquisition instruments, and the MS sensors themselves (Fig. 4.8). As depicted in the diagram, MS data captured by the sensors are transmitted to the acquisition instruments, which simultaneously collect information from the three sensors installed within a given section. The data are then relayed to the system control center via optical fibers. This particular configuration utilizes six unidirectional sensors deployed across two monitoring sections, with

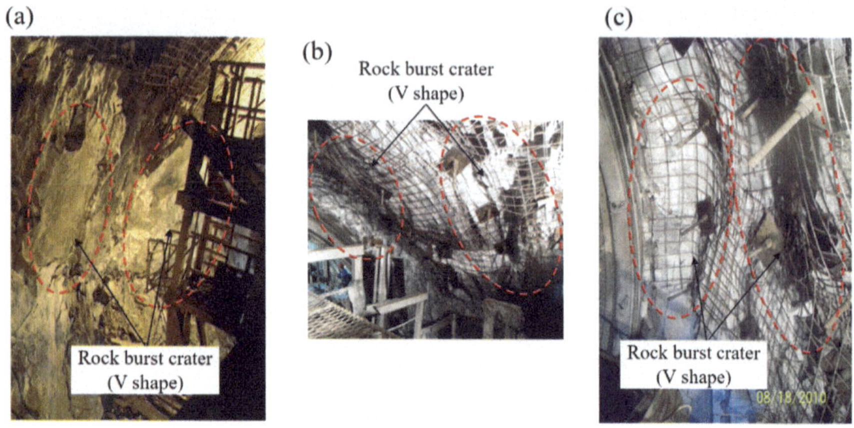

Fig. 4.7 Photos of rockbursts occurring in the Jinping tunnel. **a** 7–11 June 2010, **b** 3–9 July 2010, and **c** 16–21 August 2010. (Chen et al. 2011)

the layout within each section illustrated in Fig. 4.8. During excavation at working face 1, the six sensors are positioned in Sects. 4.1 and 4.2. When the tunnel boring machine advances 40 m to working face 2, the three sensors originally in Sect. 4.1 are retrieved and reinstalled in Sect. 4.3 (designated R3-C, R3-R, and R3-L). This iterative relocation process allows the six-sensor array to advance progressively with the tunneling activity, maintaining a consistently short distance between the sensors and the working face.

Previous research indicates that a substantial number of MS events occur in the vicinity of the working face during excavation (Feng et al. 2013). Many of these events are thus located in close proximity to the sensors, rendering S-wave arrival picking challenging due to interference from the preceding P-wave. Relying solely on P-wave arrivals for source localization. Conversely, incorporating even a single reliable S-wave arrival into the localization procedure substantially enhances accuracy (Kennett and Engdahl 1991; Wang et al. 2022; Zhang et al. 2021).

4.4.2 IPDI Analysis of an Example Rockburst Signal

To illustrate the practical application of the IPDI_S method in an engineering context, a rockburst event that occurred on June 11, 2010, at 03:14:00 is analyzed as a case study. During this event, all six sensors across the two monitoring sections were triggered, and the resulting waveforms are presented in Fig. 4.9. Compared to the simulated signals used earlier, these field waveforms exhibit considerable noise contamination. Furthermore, the overlap between P- and S-waves is significantly more pronounced due to the complexity of the real MS source and sensor ringing effects. As a result, only the S-wave arrivals for signals recorded by the three sensors

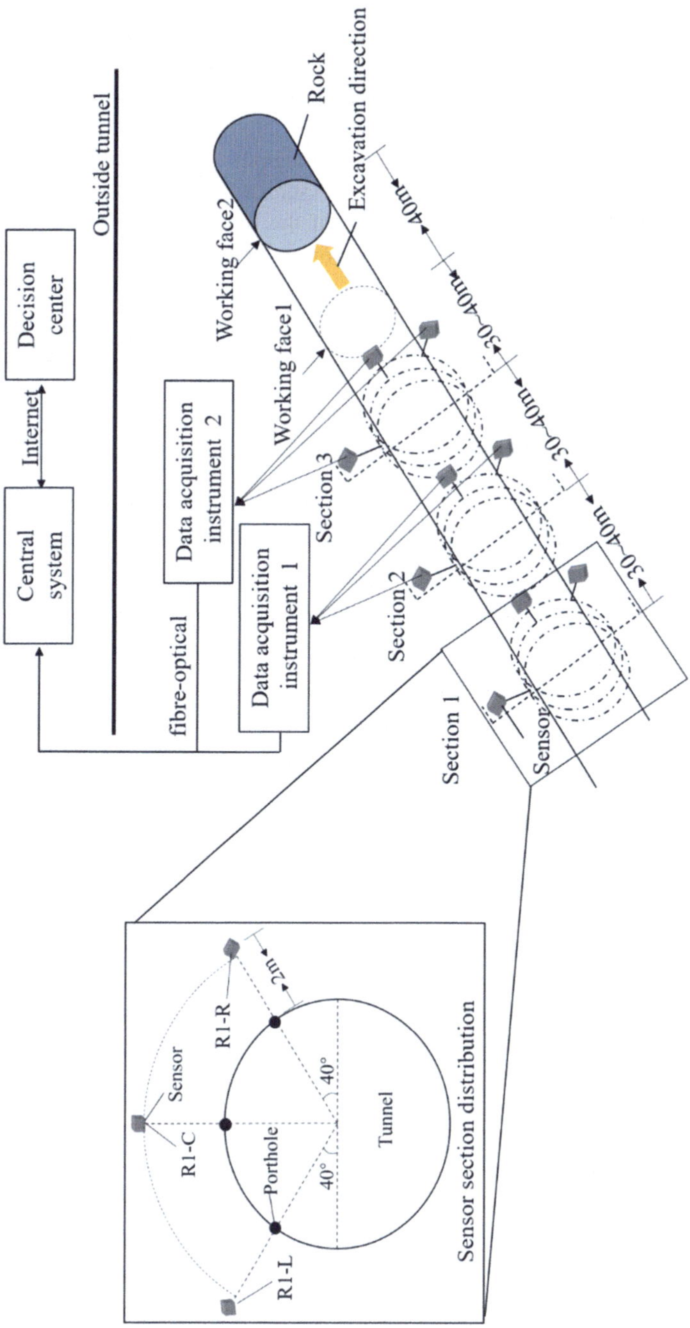

Fig. 4.8 Layout of the MS sensors in the Jinping tunnel

farther from the source (R6-C, R6-R, and R6-L) could be roughly identified through manual inspection.

In Fig. 4.9, manually picked S-wave arrivals are indicated by blue line segments, while red lines denote P-wave arrival times (actual values are listed in Table 4.1). For the sensors located closer to the source (R7-C, R7-R, and R7-L), manual identification of S-wave arrivals proved too challenging. To demonstrate the IPDI method's capability, analyses were first performed on the signals recorded at sensor R6-C (where manual S-wave picking was feasible) and sensor R7-R (where it was not).

As shown in Fig. 4.9, the triggering order of the sensors is as follows: R7-C, R7-R, R7-L, R6-C, R6-R, and R6-L. Accordingly, the corresponding signals are

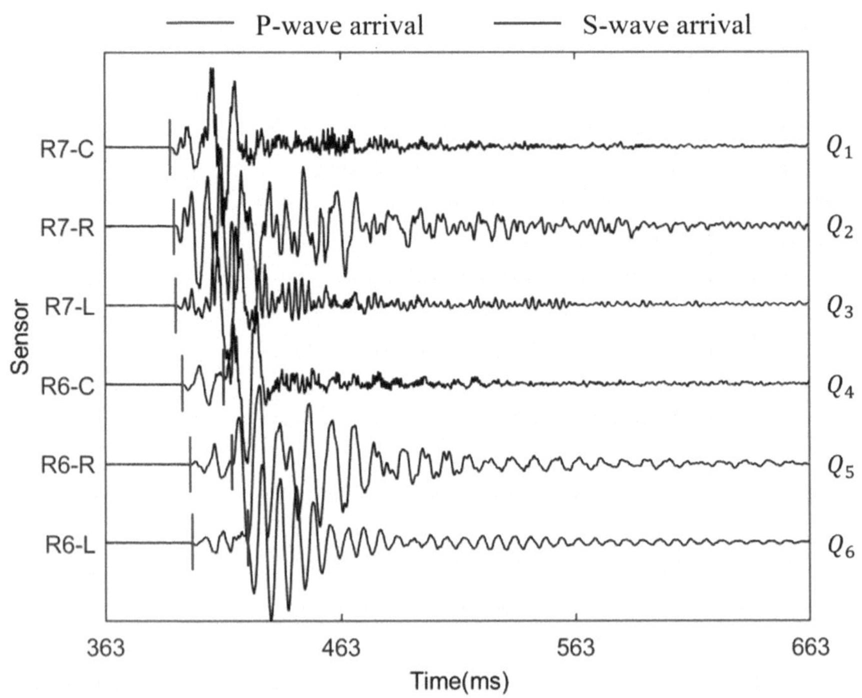

Fig. 4.9 Original signals of the rockburst that occurred on June 11th at 03:14:00

Table 4.1 Arrival times associated with the rockburst event that occurred on 2010 June 11th at 03:14:00

Wave type	Sensor					
	R7-C	R7-R	R7-L	R6-C	R6-R	R6-L
P-wave: manually picked time (ms)	389.855	391.022	392.189	396.777	397.610	398.443
S-wave: manually picked time (ms)	–	–	–	412.944	416.111	420.943
S-wave: IPDI-picked time (ms)	–	403.444	405.277	412.944	415.611	420.443

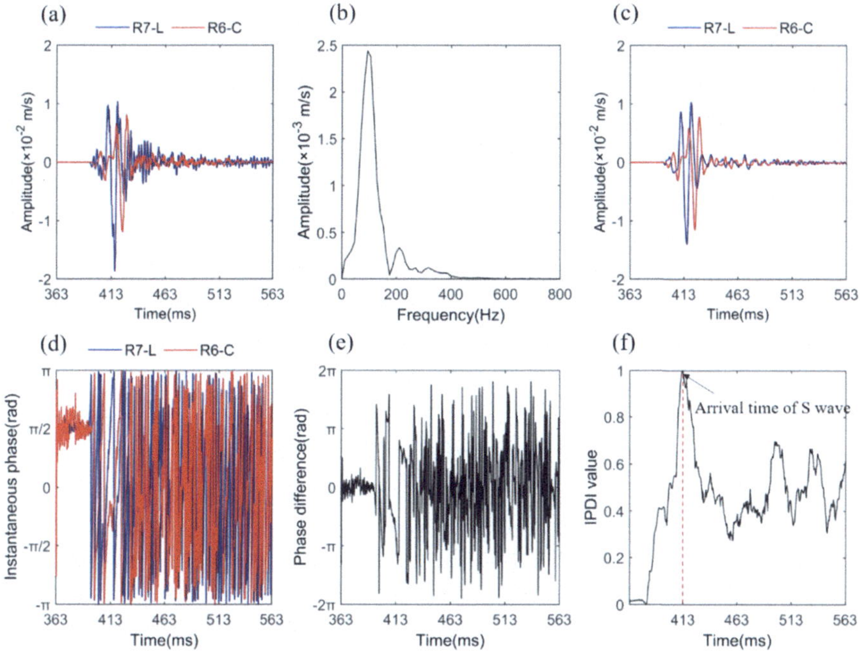

Fig. 4.10 IPDI analysis of rockburst-generated signals. **a** Displays the signals captured by sensors R6-C and R7-L. The frequency spectrum corresponding to waveform Q_4 is presented in (**b**). **c** Reconstructed signals derived from the two waveforms by EEMD. The instantaneous phases of these two reconstructed signals are plotted in (**d**), while their instantaneous phase difference is depicted in panel (**e**). Finally, the resulting IPDI curve is provided in panel (**f**).

sequentially labeled Q_i with $(i = 1, 2..., 6)$. To identify the S-wave arrival time of Q_4 (the signal recorded by sensor R6-C), the signal Q_3 from sensor R7-L is also needed. The waveforms of both signals are displayed together in Fig. 4.10a.

Figure 4.10b presents the frequency spectrum of signal Q_4, from which the dominant frequency is determined to be 93 Hz. Given the P-wave and S-wave velocities of 6154 m/s and 3059 m/s, respectively, the corresponding S-wave wavelength is calculated as 32.89 m, and its period is 10.75 ms. Based on the sensor excitation times, 392.189 ms for R7-L and 396.777 ms for R6-C, and the P-wave velocity, P is computed as 28.23 m using Eq. 4.2. Since P is less than the S-wave wavelength, the IPDI_S method is applicable for extracting the S-wave arrival time of Q_4.

Both signals Q_4 and Q_3 are first decomposed and reconstructed using EEMD, and the resulting reconstructed waveforms are shown in Fig. 4.10c. Their instantaneous phases are then obtained via HT, as illustrated in Fig. 4.10d. The instantaneous phase difference between the two signals is plotted in Fig. 4.10e. Finally, the IPDI_S curve is computed, as shown in Fig. 4.10f. The time corresponding to the maximum of this curve is 412.944 ms, which is taken as the S-wave arrival time of signal Q_4. This value exhibits no deviation from the manually picked S-wave arrival time indicated in Fig. 4.9.

The IPDI_S method was subsequently applied to identify the S-wave arrival times for the signals recorded at sensors R7-R, R7-L, R6-R, and R6-L. The corresponding waveforms and computed IPDI_S curves are presented in Fig. 4.11, permitting clear determination of the S-wave onsets. The resulting arrival times are summarized in Table 4.1. Notably, the IPDI_S method successfully identified S-wave arrivals in near-field sensor signals (e.g., R7-R and R7-L), where manual recognition proved infeasible. Moreover, for far-field sensors (R6-C, R6-R, and R6-L), where P- and S-wave overlap is less pronounced, the IPDI-derived arrival times exhibited strong consistency with the manually picked values.

4.4.3 Assessing the Accuracy of the IPDI Method

The preceding analysis of the rockburst event that occurred at 03:14 on June 11, 2010, demonstrates that the IPDI_S method is capable of identifying S-wave arrival times in signals where manual picking is infeasible. However, because the exact wave velocities and propagation paths in practical settings are unknown, the accuracy of the IPDI-selected results cannot be assessed through direct comparison with theoretical values. Instead, the method's performance is evaluated by three alternative methods, summarized in Table 4.2 and described below.

Scenario 1: A dataset comprising 131 MS events recorded between June 4 and June 14 was selected. For these events, the S-wave arrival times could be identified manually with relative ease. The manually picked values were then compared with those obtained using the IPDI_S method to assess consistency.

Scenario 2: This scenario focused on eight rockburst events that occurred during the same period. These events were categorized into two groups: those for which at least one S-wave arrival time could be manually identified (six events) and those for which no S-wave arrivals were manually discernible (the remaining two events). The IPDI method was applied to all eight events to estimate S-wave arrival times, which were then used to locate the rockbursts. The resulting locations were compared with the actual rockburst positions to evaluate the method's accuracy.

Scenario 3: The IPDI_S method was applied to 73 MS events for which manual S-wave arrival picking was not possible. The performance of the method was assessed by comparing the event locations derived using both P- and S-wave arrival times (with S-wave picks provided by IPDI_S) against those obtained using only P-wave arrivals.

(1) *Scenario 1*

The discrepancies between the S-wave arrival times derived from the IPDI_S method and those obtained through manual identification were analyzed statistically, with

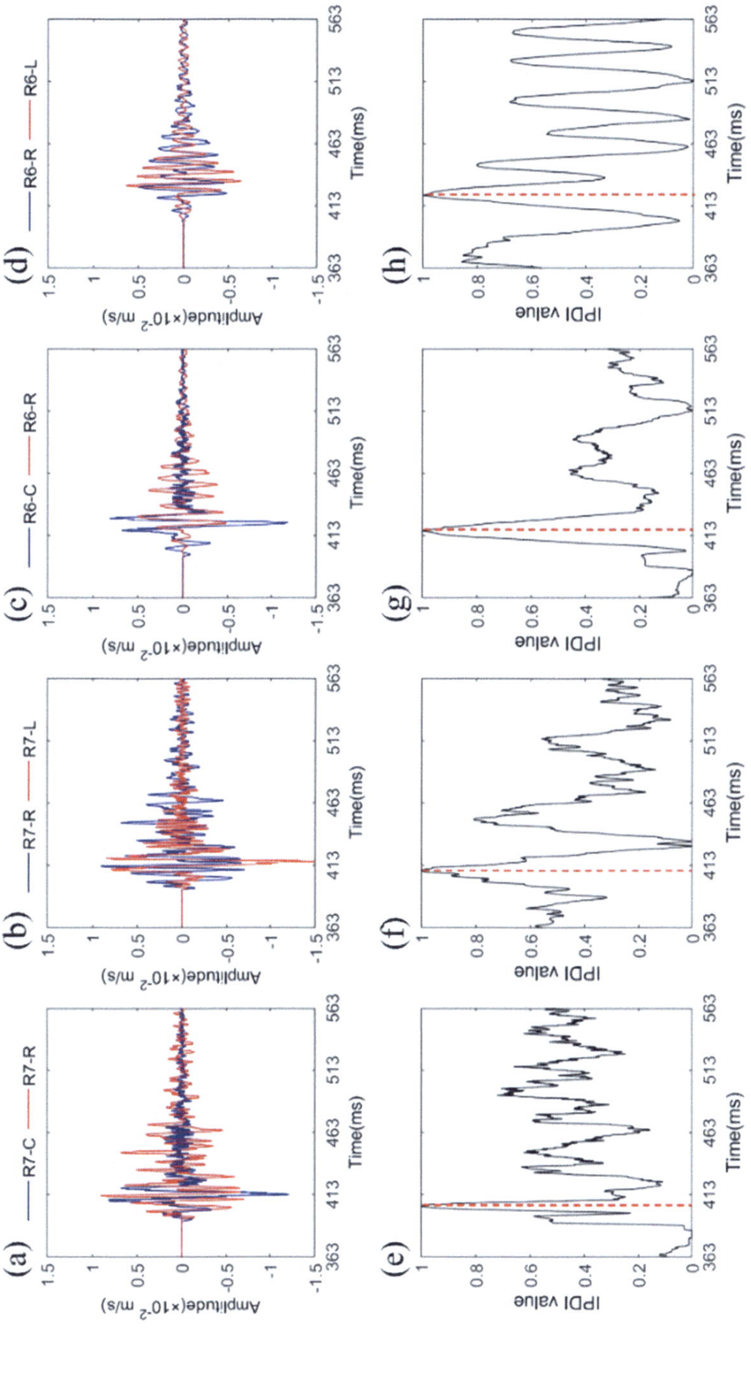

Fig. 4.11 IPDI analysis applied to additional signal pairs recorded during the rockburst event that occurred on June 11th at 03:14:00. **a**–**d** Depict the waveforms captured by specific sensor combinations: **a** R7-C and R7-R, **b** R7-R and R7-L, **c** R6-C and R6-R, and **d** R6-R and R6-L. The corresponding IPDI curves derived from these signals are displayed in (**e**)–(**h**) for sensors R7-R, R7-L, R6-R, and R6-L, respectively

Table 4.2 Details of the test scenarios

Scenario	Input data used	Quantities compared
1	131 signals that S-wave arrival times could be determined manually	Manually picked S-wave arrival times and those obtained via the IPDI method
2	8 rockburst events with known locations	Event locations derived from the IPDI method versus actual rockburst positions
3	73 microseismic events whose S-wave arrival times could not be manually picked	Event locations derived from the IPDI method and those based solely on P-wave arrival times

the results presented in Fig. 4.12. It can be observed that 71.8% of the differences are within 1 ms, and 26.0% are within 0.5 ms. Only a small proportion of signals exhibit deviations exceeding 1.5 ms. These findings indicate that the IPDI_S method provides a reliable means of picking S-wave arrival times in practical engineering applications.

(2) *Scenario 2*

In this scenario, a hierarchical particle swarm optimization algorithm (Chen et al. 2009) is employed to locate MS and rockburst events. The actual locations of the eight rockburst events considered in scenario 2 are shown in Fig. 4.13a. Among these, events 1 through 6 each had at least one manually identifiable S-wave arrival, while no S-wave arrivals could be manually identified for events 7 and 8. Figure 4.13b presents the location results for events 1–6 based on manually picked S-wave arrivals and for events 7 and 8 using only P-wave arrivals. For comparison, Fig. 4.13c shows the location results obtained when S-wave arrival times were derived using the IPDI_S method for all eight events.

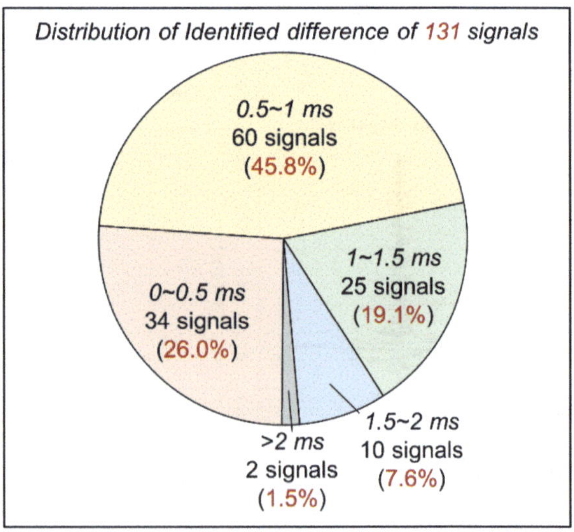

Fig. 4.12 Distribution of differences between manually picked and IPDI-derived S-wave arrival times

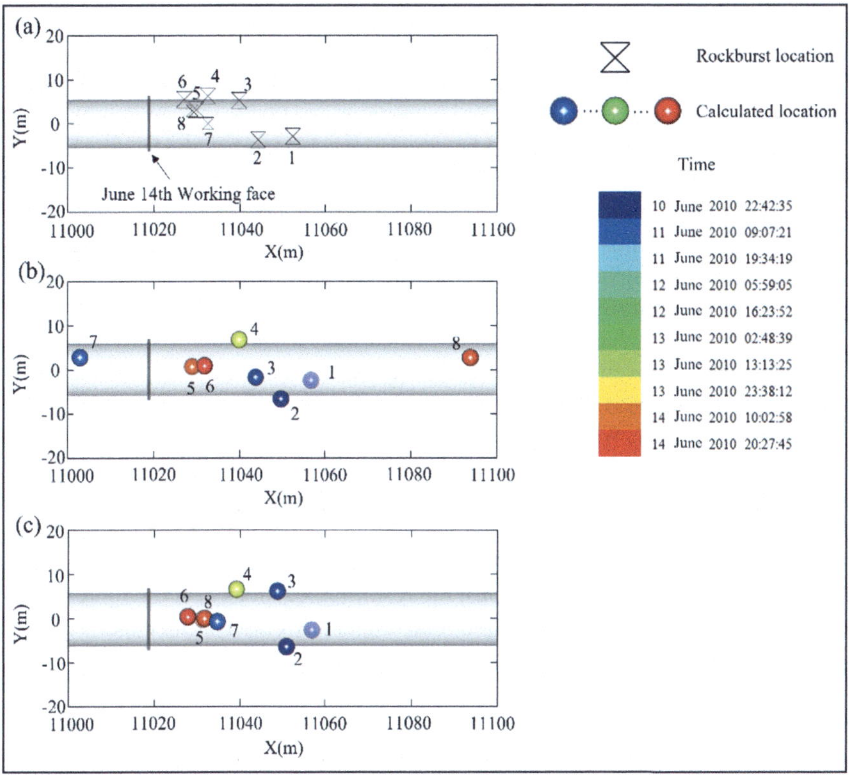

Fig. 4.13 Locations of the eight rockburst events in Scenario 2: **a** actual rockburst positions, **b** locations derived from manually picked arrival times, and **c** locations derived from IPDI-picked arrival times

A comparison of Figs. 4.13b, c reveals that for events 1–6, the locations obtained using manually picked and IPDI-derived S-wave arrivals are highly consistent with each other and with the actual rockburst positions. For events 7 and 8, however, relying solely on P-wave arrivals resulted in substantial localization errors (Fig. 4.13b). In contrast, incorporating IPDI-identified S-wave arrivals markedly improved the accuracy of the locations for these two events (Fig. 4.13c).

These findings underscore the critical role of S-wave arrival times in enhancing location accuracy, corroborating earlier studies (Kennett and Engdahl 1991; Wang et al. 2022). Furthermore, they demonstrate that the IPDI_S method reliably extracts S-wave arrivals from MS monitoring data, thereby significantly improving source location performance.

(3) *Scenario 3*

To further evaluate both the accuracy of the IPDI_S method and its influence on localization performance, the method was applied to 73 MS events for which manual identification of S-wave arrival times was not feasible. Figure 4.14a displays the locations of these events obtained using only P-wave arrivals. Subsequently, S-wave arrival times were estimated using the IPDI_S method, and the events were relocated; the results are shown in Fig. 4.14b. When compared with the known rockburst locations, the localizations based solely on P-wave arrivals appear highly scattered, with only a handful clustered near the actual rockburst positions. In contrast, the locations incorporating IPDI-derived S-wave arrivals exhibit markedly reduced dispersion, with a greater number of events concentrated around the rockburst zones.

Examining the events chronologically, with colors in Fig. 4.14 indicating occurrence time, reveals that the IPDI_S method substantially enhances location accuracy by successfully identifying near-field S-wave arrivals. Events occurring before June 7, 2010 (depicted in blue), are more dispersed than those recorded between June 7 and 14. This improvement can be attributed to the fact that prior to June 7, the monitoring sensors were deployed in sections R4 and R5, approximately 114 m from the working face, whereas after June 7, they were relocated to sections R6 and R7, approximately 146 m away. Thus, the earlier period corresponds to near-field monitoring conditions, where S-wave onsets are heavily influenced by preceding P-wave energy, complicating accurate arrival picking and leading to greater localization scatter. By applying the IPDI_S method to extract S-wave arrivals under such challenging near-field conditions, the resulting locations become more tightly clustered around the rockburst positions and align more closely with expected physical patterns. These findings demonstrate that the IPDI_S method offers significant advantages for processing near-field MS monitoring data.

The three evaluation scenarios considered above demonstrate that the IPDI_S method reliably identifies S-wave arrival times in MS monitoring, performing effectively regardless of signal clarity. As a result, it enables more accurate locations of MS events, thereby enhancing the reliability of subsequent rockburst warnings.

4.5 Discussion

The simulation and field investigations demonstrate that the IPDI_S method accurately determines S-wave arrival times for MS events in tunnels, thereby enhancing event location precision. Nevertheless, several aspects merit further investigation and discussion.

(1) *Parameter sensitivity*

The IPDI_S method involves two critical parameters: the integration interval, corresponding to the S-wave period T, and the targeted phase difference ω_e. To examine the influence of the integration interval, the data used in the analysis presented in Fig. 4.5 were reprocessed with varying integration lengths, as shown in Fig. 4.15a. Increasing the interval reduces noise in the IPDI_S curve and enhances the prominence of the primary peak. As the interval expands from $0.5T$ to $2T$, the amplitude of secondary fluctuations near the maximum diminishes. However, extending the

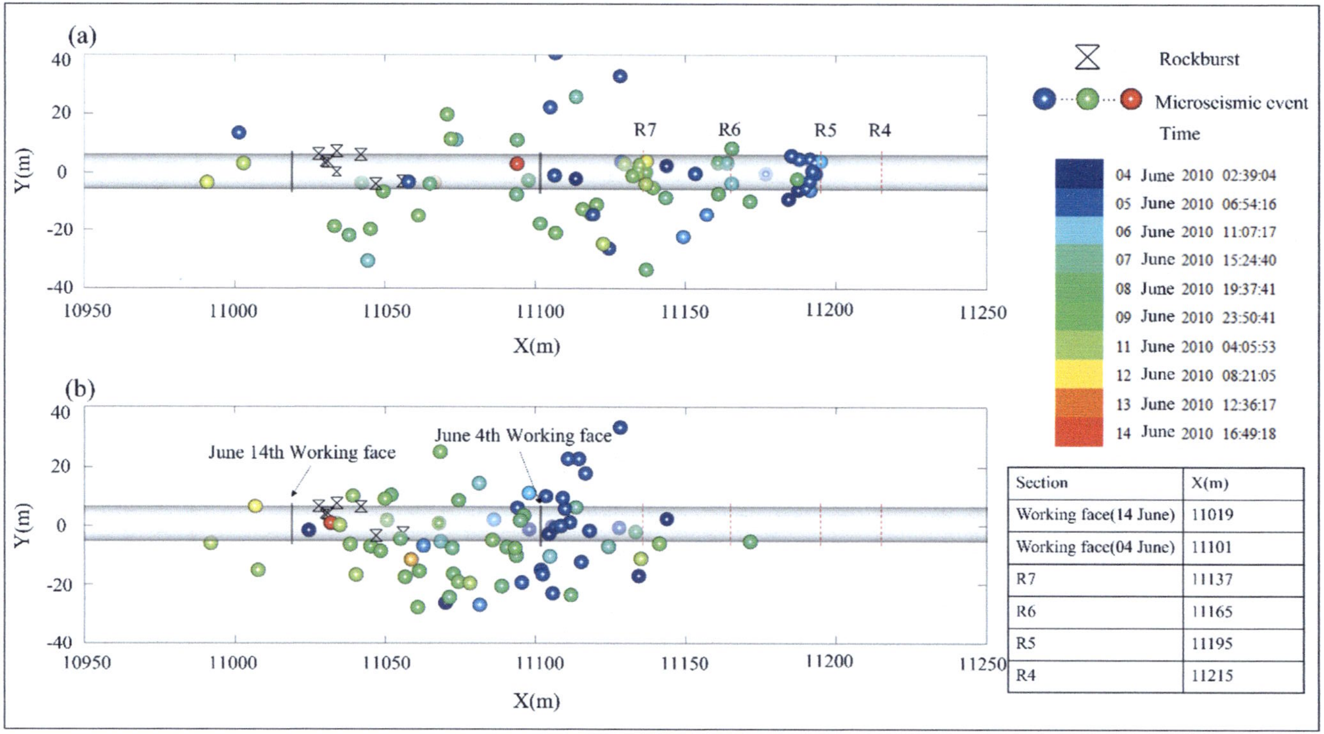

Section	X(m)
Working face(14 June)	11019
Working face(04 June)	11101
R7	11137
R6	11165
R5	11195
R4	11215

Fig. 4.14 Locations of the 73 MS events considered in scenario 3. The location predictions shown are based on **a** the P-wave arrival time and **b** the P- and S-wave arrival times

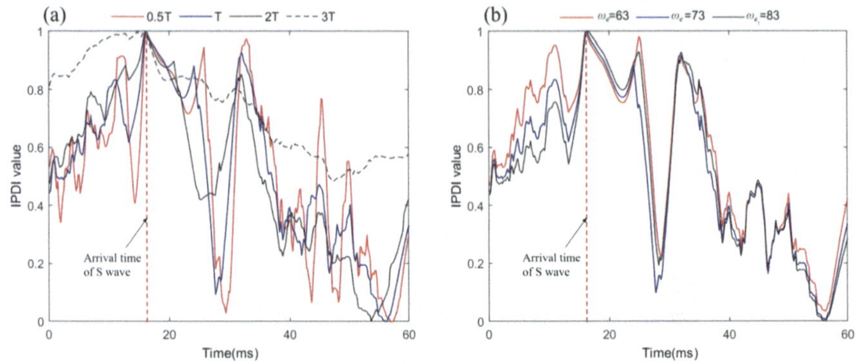

Fig. 4.15. Results of two parameter-sensitivity analyses. The graphs shown correspond to the IPDI curves produced when different **a** integration intervals and **b** targeted phase differences are used in the analysis

interval excessively, such as to $3T$, leads to a loss of resolution, increasing the risk of significant picking errors.

Regarding the targeted phase difference, the IPDI_S method exhibits a degree of robustness, as illustrated in Fig. 4.15b. When ω_e is varied by $\pm 10°$ from its true value of 73°, the location of the maximum peak remains virtually unchanged. This indicates that the method tolerates an error of approximately $\pm 10°$ in the estimation of ω_e. Since ω_e is derived from P- and S-wave velocities, this tolerance effectively accommodates a certain level of uncertainty in velocity estimation.

(2) *Applicability of the IPDI method*

The IPDI_S method has proven effective for MS monitoring in tunnels, as the sensor configurations typically employed satisfy its underlying requirements. Specifically, the method necessitates that P be less than L_S, meaning that the spacing between adjacent sensors must not be excessive. In cases where the sensor separation is large such that $P > L_s$, the targeted phase difference α_e can still be determined using the following expression:

$$\omega_e = \text{REM}\left(\frac{P}{L_s}\right) \times 2\pi \tag{4.7}$$

where "REM" is the remainder function. However, the calculated ω_e value differs significantly from the true value when $P > L_s$. The IPDI_S method is predicated on the assumption that signals received by adjacent sensors share the same dominant frequency. However, empirical evidence indicates that the dominant frequency of a stress wave, as obtained through Fourier transform, evolves during propagation (Zhou et al. 2016; Sivaram and Gupta 2022). This phenomenon has been further corroborated by Xiao et al. (2015) through statistical analysis of numerous MS events. Consequently, when sensors are spaced far apart, the dominant frequencies of their recorded signals may differ substantially, leading to significant inaccuracies in S-wave arrival times estimated via the IPDI_S method. To mitigate errors arising from frequency dispersion, adjacent sensors should be positioned as closely as possible. In applications such as mining or oilfield monitoring, where intersensor distances can extend to hundreds or even thousands of meters, frequency dispersion becomes particularly pronounced. Thus, the IPDI_S method is best suited for tunnel engineering, where sensor spacing is inherently limited.

(3) *Limitations of the IPDI method*

Several aspects of the IPDI_S method warrant further investigation. First, boundary effects inherent in the ensemble EEMD procedure can compromise the accuracy of peak identification in the IPDI curve. The spurious peaks observed at the ends of these curves are attributable to such effects. A potential strategy involves artificially extending the original signal by appending replicated end segments, a strategy that merits future exploration.

When the propagation paths from the source to two adjacent sensors are of equal length, P becomes zero, causing both sensors to be triggered simultaneously. In this scenario, the targeted phase difference ω_e for both P- and S-waves vanishes, rendering the IPDI_S method ineffective for S-wave arrival identification. To address this issue, it is advisable to introduce a small axial offset between sensors within the same section, thereby ensuring a nonzero P.

4.6 Conclusion

This chapter proposes the IPDI_S method for picking S-wave arrival times in tunnel MS monitoring. The method leverages the instantaneous phase difference between signals recorded by two adjacent sensors to identify S-wave onsets. Central to this technique is the introduction of the instantaneous phase difference intensity (IPDI), which accentuates the S-wave arrival. The implementation involves several signal processing steps: fast Fourier transform (FFT), ensemble empirical mode decomposition (EEMD), Hilbert transformation (HT), and construction of the IPDI curve, from which the S-wave first break is determined by identifying prominent peaks.

The effectiveness of the IPDI_S method was first evaluated using a range of synthetic signals. Comparisons between theoretical and identified arrival times demonstrated that the method reliably detects S-wave arrivals, even when the initial portion of the S-wave overlaps with the P-wave. Subsequently, the method was applied to field data acquired from the deeply buried tunnels of the Jinping II hydropower station. Analysis of numerous MS signals for which manual S-wave picking was feasible showed that over 90% of the IPDI-derived arrival times agreed with manual picks to within 1.5 ms. Moreover, the method successfully identified S-wave arrivals in waveforms where manual identification proved impractical. In further validation, the IPDI-picked arrival times were incorporated into the source location using a particle swarm optimization algorithm. The resulting locations for both rockburst and MS events exhibited markedly improved accuracy compared to those based solely on P-wave data.

References

Carcione JM (1999) Staggered mesh for the anisotropic and viscoelastic wave equation. Geophysics 64(6):1863–1866

Chen BR, Feng XT, Li SL, Yuan JP, Xu SC (2009) Microseism source location with hierarchical strategy based on particle swarm optimization. Chin J Rock Mech Eng 28(4):740–749

Chen BR, Feng XT, Zeng X, Xiao YX, Zhang Z, Ming H, Feng G (2011) Real-time microseismic monitoring and its characteristic analysis during TBM tunneling in deep-buried tunnel. Chin J Rock Mech Eng 30(2):275–283

Feng GL, Feng XT, Chen BR, Xiao YX (2017) Performance and feasibility analysis of two microseismic location methods used in tunnel engineering. Tunnell Underground Space Technol 63:183–193

Feng GL, Feng XT, Chen BR, Xiao YX, Zhao ZN (2019a) Effects of structural planes on the microseismicity associated with rockburst development processes in deep tunnels of the Jinping-II Hydropower Station, China. Tunnell Underground Space Technol 84:273–280

Feng GL, Feng XT, Xiao YX, Yao ZB, Hu L, Niu WJ, Li T (2019b) Characteristic microseismicity during the development process of intermittent rockburst in a deep railway tunnel. Int J Rock Mech Mining Sci 124:104135

Feng XT, Chen BR, Zhang CQ, Li SJ, Wu SY (2013) Mechanism, warning and dynamic control of rockburst development processes. China Social Sciences Publishing House

Frid M, Kamai R (2020) An analytical approach for estimating the spectral P/S ratio within ground motions. Comput Geotech 119:103379

Kennett BLN, Engdahl ER (1991) Traveltimes for global earthquake location and phase identification. Geophys J Int 105(2):429–465

Kuperkoch L, Meier T, Brüstle A, Lee J, Friederich W (2012) Automated determination of S-phase arrival times using autoregressive prediction: application to local and regional distances. Geophys J Int 188(2):687–702

Li N, Wang E, Ge M, Sun Z, Li B (2014) A method for identifying outlier signals for microseismic event based on arrival time difference. Chin J Rock Mech Eng 33(8):1654–1661

Liu F, Tang CA, Ma T, Tang L (2019) Characterizing rockbursts along a structural plane in a tunnel of the Hanjiang-to-Weihe river diversion project by microseismic monitoring. Rock Mech Rock Eng 52:1835–1856

Liu L, Li SJ, Xiao YX, Chen SJ, Wang ZF, Feng GL, Wang Y (2023) An inverted heterogeneous velocity model for microseismic source location in deep buried tunnels. Rock Mech Rock Eng, pp 1–26

Niu WJ, Feng XT, Xiao YX, Feng GL, Yao ZB, Hu L (2021) Identification of potential high-stress hazards in deep-buried hard rock tunnel based on microseismic information: a case study. Bull Eng Geol Environ 80:1265–1285

Sivaram K, Gupta S (2022) Frequency-dependent attenuation characteristics of coda and body waves in the Kumaun Himalaya: implications for Regional Geology and Seismic Hazards. Pure Appl Geophys 179(3):949–972

Wang G, Luo Y, Gong H, Liu T, Li X, Song L (2023) Investigations on the macro-meso mechanical properties and energy dissipation mechanism of granite shear fracture under dynamic disturbance. Int J Numer Anal Methods Geomech

Wang Z, Li J, Li B (2022) Source location mechanism of microseismic monitoring using PS waves and its effect analysis. DYNA-Ingeniería e Industria 97(1)

Wu Z, Huang NE (2009) Ensemble empirical mode decomposition: a noise-assisted data analysis method. Adv Adapt Data Analysis 1(01):1–41

Xiao Y, Chen S, Feng G et al (2025) Picking P-wave arrival times from microseismic signals in tunnels with low signal-to-noise ratios without noise filtering. Soil Dyn Earthquake Eng 192:109288

Xiao Y, Chen S, Liu L et al (2024) Determining S-wave arrival times for microseismic monitoring in tunnels based on instantaneous phase difference of sensors. Rock Mech Rock Eng 57(3):1781–1797

Xiao YX, Feng XT, Chen BR, Feng GL (2015) Evolution of frequency spectrum during instant rockbursts in deep inoculation tunnel. Rock Soil Mech 36(4):1127–1134

Zhang W, Feng XT, Bi X, Yao ZB, Xiao YX, Hu L, Niu WJ, Feng GL (2021) An arrival time picker for microseismic rock fracturing waveforms and its quality control for automatic localization in tunnels. Comput Geotech 135:104175

Zhou J, Lu W, Yan P, Chen M, Wang G (2016) Frequency-dependent attenuation of blasting vibration waves. Rock Mech Rock Eng 49:4061–4072

Zhu QJ, Jiang FX, Wang CW, Liu Y, Wang JJ, Yu ZX (2013) Automated microseismic event arrival picking and multi channel recognition and location. J China Coal Soc 38(3):397–403

Chapter 5
Absolute Location Method for MS Event Clusters in Deep Tunnels Based on Coda Wave Interference

5.1 Prologue

MS monitoring technology enables real-time acquisition and analysis of signals generated by fracturing in surrounding rock, thereby allowing the evolution of rock damage to be tracked. From these signals, multiple MS parameters are extracted to facilitate rockburst early warning (Li et al. 2023; Xiao et al. 2026). As a fundamental prerequisite for deriving such parameters, the accurate location of MS events directly influences the reliability of subsequent warning assessments (Wamriew et al. 2022; Xiao et al. 2024, 2025). Thus, precise event location constitutes a critical component of MS-based analysis.

The most commonly employed method for source location is the traditional traveltime difference method, originally proposed by Geiger (1912). This technique requires the manual or automated picking of P- and S-wave arrival times for each recorded signal, followed by the formulation of traveltime difference equations in which the hypocenter coordinates and origin time are treated as unknown parameters. To solve these nonlinear functions under varying engineering conditions, researchers have adopted a range of optimization algorithms (Chen et al. 2023). For example, Chen et al. (2009) developed a layered location method based on the particle swarm optimization (PSO) algorithm, which demonstrated satisfactory performance in mining applications. Zhang et al. (2023) introduced an artificial bee colony algorithm tailored for large underground powerhouse projects. More specifically, for tunneling environments, Feng et al. (2015) proposed a method that models the velocity structure between the source and sensors as a series of vertical layers and integrates the PSO algorithm to enhance location accuracy. Despite these advancements, such methods generally assume a homogeneous velocity field for the surrounding rock, thereby neglecting the inherent complexity of the actual velocity structure. Consequently, this simplification inevitably introduces errors when these approaches are applied in complex engineering settings.

© Tongji University Press Co., Ltd 2026

Y. Xiao et al., *Near-Field Microseismic Data Processing in Deep Tunnels*,
https://doi.org/10.1007/978-981-92-3208-6_5

To overcome the inherent errors introduced by such simplifications, it is necessary to replace the homogeneous velocity model with a heterogeneous one (Peng and Wang 2019). To determine the MS location and origin time using traveltime differences, obtaining the minimum traveltime of the waves is essential. Commonly adopted strategies for this purpose include ray-tracing methods and traveltime field methods. Ray tracing, grounded in high-frequency approximation, iteratively solves the ray equations to compute both propagation paths and traveltimes (Ding et al. 2018). Traveltime field methods compute full-field traveltimes by solving the Eikonal equation via techniques such as the fast marching method (FMM) (Jiang et al. 2021; Zhang et al. 2022) or the fast sweeping method. Nevertheless, these methods rely on an accurately predefined heterogeneous velocity model to achieve precise location. Moreover, they are typically applied to individual MS events. In practice, however, MS events frequently occur in clusters during the incubation phase of rockbursts, reflecting the continuous nature of rock fracturing and stress release. Thus, there is a need for a location method that does not depend on a preestablished velocity model and is capable of handling clusters of MS events.

Double-difference location and tomography techniques offer a coupled approach to locating event clusters while simultaneously reconstructing the velocity model. The double-difference location concept was first introduced by Waldhauser and Ellsworth (2000) and subsequently extended by Zhang and Thurber (2003). The main idea involves computing double differences between observed and calculated absolute arrival time differences for pairs of events, thereby allowing for source location independent of a known velocity model. This method has been widely applied in aftershock sequence analysis, fault-zone structure characterization, and volcanic activity monitoring (Waldhauser 2009; Barajas et al. 2022). The performance of the double-difference method is strongly influenced by the density of the station network, making it particularly suitable for regions with dense station coverage. However, in deeply buried underground projects such as tunnels, the number of installed MS sensors is typically limited to six to nine (He et al. 2025) and may be even fewer if some sensors become inoperative. Under such conditions, both double-difference and conventional traveltime difference methods become infeasible, as the number of available equations is insufficient to constrain the number of unknown parameters.

Coda wave interferometry (CWI) offers a robust solution to the challenges outlined above. Unlike traveltime-based or double-difference methods, CWI estimates the spatial separation between seismic sources by analyzing the cross-correlation of coda waves from paired events (Zhao and Curtis 2019). Due to the multiple scattered phases contained within coda waves, their effective path lengths substantially exceed those of direct waves, rendering them highly sensitive to minor variations in source location. Consequently, CWI enables high-precision relative event location even with sparse sensor arrays. This technique has been successfully applied in large-scale geophysical studies (Snieder and Vrijlandt 2005). However, as it provides only relative locations of event clusters, it does not fully satisfy the requirements for rockburst early warning.

This chapter presents an improved method based on CWI that achieves the transformation from relative to absolute location of MS event clusters by integrating reference events with known source coordinates into the analysis. The feasibility of applying joint CWI analysis to both blasting and rockburst signals is evaluated through examination of their respective coda wave characteristics. A tunnel scenario is subsequently designed to preliminarily assess the localization performance of the proposed method. The approach is then applied to the Bayu Tunnel case, where it is used to locate blasting signals, rockburst signals, and MS event clusters to evaluate its performance. Finally, a sensitivity analysis of key CWI parameters is performed.

5.2 The Characteristics of the Coda Wave in the Tunnel

5.2.1 Background

The Bayu tunnel is situated in the Sangga downstream section of the Zangnan Valley, a region defined by typical alpine gorge topography. The tunnel elevation spans from 3260 to 5500 m, corresponding to a vertical relief of approximately 2,300 m. The main tunnel extends 13,073 m in length, of which 12,242 m, accounting for 94% of the total, is predicted to be susceptible to rockbursts. This includes 4,106 m classified as mild rockburst zones and 1,158 m as moderate rockburst zones. Detailed geological descriptions of the Bayu Tunnel can be found in Wang et al. (2021) and Li et al. (2022). The regional stress field is governed by strike-slip faulting, with the maximum principal stress oriented between N20°W and N30°E, reflecting substantial horizontal tectonic forces. Figure 5.1 presents photographs of four distinct rockburst events, illustrating the severe damage they cause. These observations underscore the critical necessity of implementing microseismic monitoring and rockburst early warning systems in the Bayu Tunnel.

The MS data utilized in this chapter were collected from the Bayu Tunnel between May 8 and May 18, 2019, using two sensor arrays comprising a total of eight sensors. These arrays were initially positioned approximately 100 and 130 m from the working face, respectively. As excavation advanced by 30 m, the array farther from the face was relocated to maintain a distance of approximately 100 m from the new working face. This rolling deployment strategy enabled continuous and efficient monitoring of MS activity throughout the excavation process (Fig. 5.1).

During this monitoring period, a total of eight blasting events and four rockburst events occurred. The number of MS events, rockburst events, and blasting events recorded by the sensors is summarized in Table 5.1. It should be noted that while all eight sensors successfully captured complete signals for both blasting and rockburst events, MS events were not consistently detected by every sensor. As a result, the total number of recorded MS signals is lower than the theoretical maximum (363 events $\times$ 8 traces).

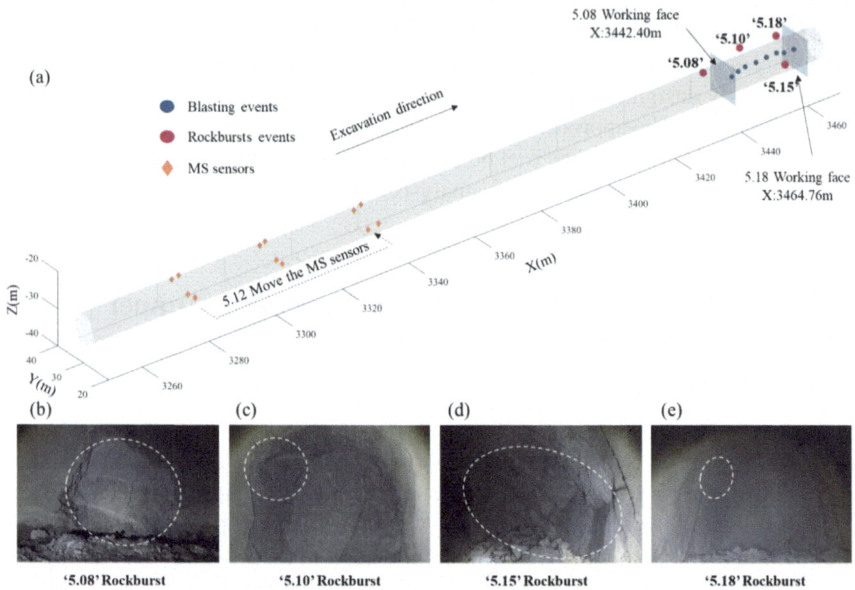

Fig. 5.1 Overview of the Bayu tunnel during the monitoring period (May 8–18, 2019). **a** Configuration of sensors, blasting sources, rockburst sources, and working face. **b–e** Show the '5.08', '5.10', '5.15', and '5.18' rockburst events

Table 5.1 The number of events and signals of MS, rockbursts, and blasting

Quantity	Blasting	Rockburst	MS
Events	8	4	363
Signal	64	32	2063

5.2.2 Determination of Coda Waves

This section presents the theoretical basis of coda waves, outlines the key assumptions underlying coda-based methods, and clarifies the distinction between coda waves and direct waves in MS signals recorded in tunnel environments.

Coda waves constitute the latter portion of a seismic signal, characterized by sustained vibrations that follow the arrival of direct waves. As illustrated in Fig. 5.2, waves initially propagate from the source, with the direct wave being the first to reach the sensor. Subsequently, coda waves undergo multiple reflections and diffractions among scatterers before being recorded.

CWI methods are based on this assumption that signals originating from different sources share identical propagation paths between scatterers when recorded at the same sensor. Although the propagation paths of direct waves vary significantly with source location, as depicted by the red and blue lines in Fig. 5.2, multiple scattered waves traverse equivalent paths before reaching the sensor, as represented by the

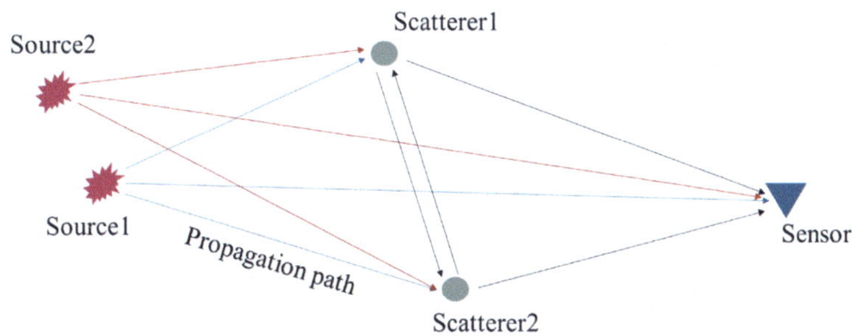

Fig. 5.2 Diagram depicting wave propagation: red and blue lines show direct-wave paths from different sources, whereas the black line represents the shared scattered path generated by scatterers

black lines. This invariance of scattering paths constitutes the theoretical basis for source location using coda waves.

A practical challenge in coda wave analysis lies in the unambiguous definition of the onset and termination of the coda segment within a signal. Currently, no universally accepted criteria exist for delineating these boundaries. Nevertheless, in geophysical applications, complete isolation of the coda segment is not essential for effective analysis; thus, the absence of precise definitions does not substantially hinder practical implementation (Snieder et al. 2002). In this study, the following operational definitions are adopted: the coda onset is set at 150 ms after the P-wave arrival, and the coda termination is determined using the established STA/LTA method (Chen et al. 2019).

5.2.3 *The Estimated Length of the Coda Wave*

Owing to the inherent difficulty in precisely measuring the full duration of coda waves, the coda lengths reported in this study are estimated using the method described in Sect. 5.2.2. These statistically derived values provide the basis for selecting appropriate coda windows in the subsequent CWI analysis.

During the monitoring period from May 8 to May 18, 2019, the sensor network in the Bayu Tunnel recorded a total of 64 blasting signals, 32 rockburst signals, and 2,063 MS signals (Table 5.1). Figure 5.3 presents the estimated coda length distributions for these signals, all of which are below 1,000 ms. Distinct clustering patterns can be observed among signal types: blasting signals predominantly exhibit coda lengths between 400 and 700 ms, while rockburst signals are mainly concentrated within the 200–500 ms range. In contrast, MS signals display a more uniform distribution across the entire observed spectrum. Notably, blasting signals consistently yield the longest coda durations among all categories.

The distributions shown in Fig. 5.3 confirm that despite considerable variability across signal types, measurable coda waves are present in all recorded events. Based on these observations, a 200 ms analysis window is adopted as optimal for coda

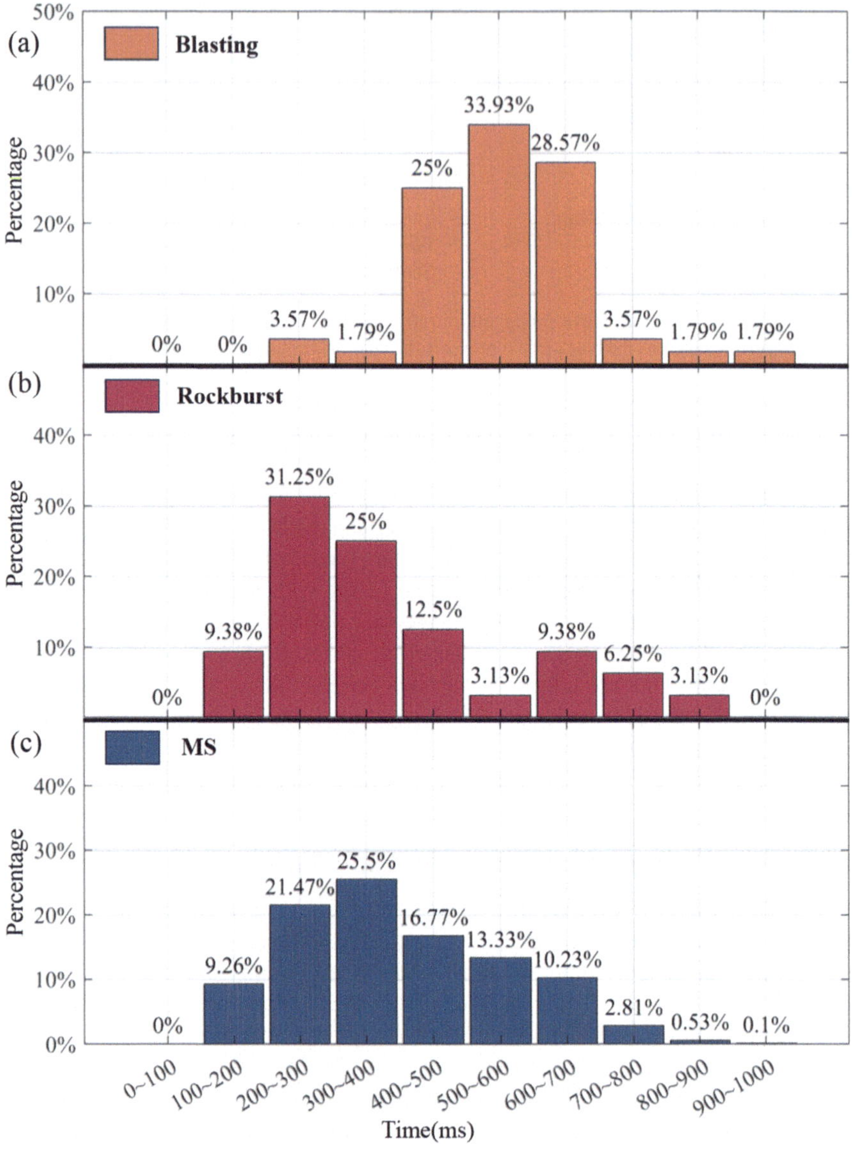

Fig. 5.3 Percentage distribution of coda wavelengths for **a** blasting signals, **b** rockburst signals, and **c** MS signals

wave analysis. This window length enables effective processing of 100% of blasting signals, 90.62% of rockburst signals, and 90.74% of MS signals. The remaining 9.38% of rockburst signals and 9.26% of MS signals, which exhibit insufficient coda lengths, are excluded from further analysis. Given their relatively small proportions, these exclusions do not materially affect the overall findings or conclusions of the study.

5.2.4 Correlation Between MS Signals and Other Signals

This section examines the correlation between the coda waves of MS signals and those generated by either blasting or rockburst events. The objective is to assess whether blasting and rockburst signals can be effectively combined with MS signals in CWI applications.

To conduct this analysis, signal pairs are constructed, each comprising one MS signal and one blasting or rockburst signal. Both signals in a given pair must originate from the same sensor. The normalized cross-correlation for each pair is then calculated according to the following expression:

$$C_{S_1S_2}(\alpha) = \frac{\int_{-\infty}^{\infty} S_1(t)S_2(t-\alpha)\mathrm{d}t}{\sqrt{\int_{-\infty}^{\infty} |S_1(t)|^2\mathrm{d}t}\sqrt{\int_{-\infty}^{\infty} |S_2(t)|^2\mathrm{d}t}} \tag{5.1}$$

where t denotes the time lag. S_1 and S_2 represent the two signals in a pair, where one is an MS signal and the other is either a blasting or a rockburst signal. $C_{S1S2}(\alpha)$ denotes the normalized cross-correlation value of S_1 and S_2 at a time lag of α.

For each signal pair, the maximum correlation coefficient is defined as the peak value of $R_{S1S2}(\alpha)$. As illustrated in Fig. 5.4, over 84% of blasting–MS pairs and more than 88% of rockburst–MS pairs exhibit correlation coefficients exceeding 0.5. The majority of blasting–MS correlations fall within the 0.70–0.85 range, while rockburst–MS correlations are primarily concentrated between 0.75 and 0.80. These results indicate a strong correlation between blasting, rockburst, and MS signals in the tunnel environment, suggesting that the three source types share comparable mechanisms. Consequently, their coda waves can be effectively employed in interferometric analysis.

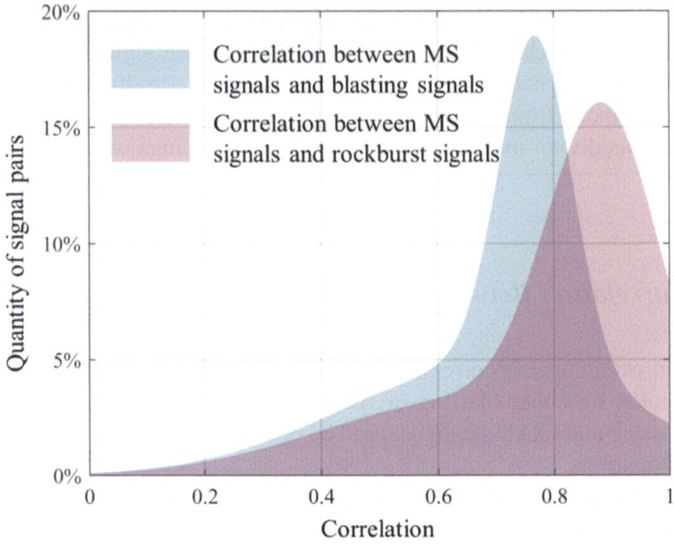

Fig. 5.4 Cross-correlation analysis of MS signals with blasting (blue) and rockburst (red) signals

5.3 Methodology

5.3.1 Coda Wave Interference

An MS signal $S_1(t)$ can be decomposed into the superposition of contributions from all wave propagation paths.

$$S_1(t) = \sum_T A_T(t) \tag{5.2}$$

where $A_T(t)$ represents the signaling component propagated along path T. The signal $S_2(t)$ can be expressed as:

$$S_2(t) = \sum_T A_T(t + \tau_T) \tag{5.3}$$

where τ_T denotes the traveltime difference along path T resulting from the different source locations.

The degree of correlation between the coda waves of $S_1(t)$ and $S_2(t)$ is determined by applying a stretching and compression technique to their cross-correlation.

$$R^{(t_1,t_2)}(\in) = \frac{\int_{t_1}^{t_2} (S_1(t) S_2(t[1+\in]))\mathrm{d}t}{\sqrt{\int_{t_1}^{t_2} \left(S_1{}^2(t)\right)\mathrm{d}t \int_{t_1}^{t_2} \left(S_2{}^2(t[1+\in])\right)\mathrm{d}t}} \tag{5.4}$$

where $R^{(t_1,t_2)}$ represents the correlation between two signals within the time interval from t_1 to t_2, and ϵ denotes the time stretch factor. The maximum value of $R^{(t_1,t_2)}$ is recorded as $R_{\max}$. Meanwhile, $R_{\max}$ can also be expressed as:

$$R_{\max} = 1 - \frac{1}{2\omega_{\mathrm{r}}^2\sigma_\tau^2} \tag{5.5}$$

where σ_τ denotes the time perturbation and ω_{r} represents the mean square angular frequency of the recorded waveform, which can be calculated as follows:

$$\omega_{\mathrm{r}}^2 = \frac{\int_{t_1}^{t_2} \dot{s}(t)\mathrm{d}t}{\int_{t_1}^{t_2} s(t)\mathrm{d}t} \tag{5.6}$$

where $\dot{s}(t)$ is the time derivative of the waveform $s(t)$. σ_τ can be determined based on Eqs. 5.5–5.6. Furthermore, σ_τ exhibits the following relationship with the source separation r.

$$\sigma_\tau^2 = \frac{1}{2v_{\mathrm{p}}^2} r^2 \tag{5.7}$$

$$\sigma_\tau^2 = \frac{1}{3v_{\mathrm{p}}^2} r^2 \tag{5.8}$$

$$\sigma_\tau^2 = \frac{\frac{6}{v_p^8} + \frac{7}{v_s^8}}{\frac{14}{v_p^6} + \frac{21}{v_s^6}} r^2 \tag{5.9}$$

Here, v_p and v_s denote the velocities of the P- and S-waves, respectively. Eqs. 5.7–5.9 describe the relationship between σ_τ and r in 2D isotropic, three-dimensional isotropic, and anisotropic media, thereby enabling the determination of r.

Minor discrepancies in coda waves can arise from variations in the velocity structure of the surrounding rock in tunnels (Fard et al. 2020; Liu et al. 2024). In the present study, such velocity perturbations are considered negligible, and the velocity field is assumed to remain temporally invariant. Moreover, prior research suggests that when the spatial separation between seismic sources is relatively large, estimated values tend to underestimate the true distances. To compensate for this bias, Robinson et al. (2011) and Snieder (2006) introduced a conditional density function into the source separation estimation.

Based on the calculated separations for each event pair, an objective function is formulated, thereby recasting the relative location problem as an optimization task (Snieder 2006). This optimization is subsequently solved using the conjugate gradient method.

$$L(e_1, e_2, \ldots, e_n) = -\sum_{k=1}^{N_{\mathrm{sensor}}} \sum_{i=1}^{n-1} \sum_{j=i+1}^{n} \ln\left[P_k\left(\breve{\delta}^{k}_{cwi,ij} | e_i, e_j\right)\right] \tag{5.10}$$

where $\delta^k_{cwi,ij} = \left\| e_i - e_j \right\|_2$ denotes the source separation between events i and j determined from the k-th sensor. $e_i = (x_i, y_i, z_i)$ is the event location. N_{sensor} represents the total number of sensors. n is the number of MS events considered for relative location. L denotes the residual of the objective function.

If the absolute locations of three events are known, the cluster can be spatially translated and rotated to transform their relative locations into absolute locations.

5.3.2 Using CWI for MS Event Cluster Location in Tunnels

The workflow for locating MS event clusters by the CWI method can be divided into four steps (Fig. 5.5).

(1) **Event Clustering**

The raw input data consist of waveforms and sensor information, including sensor names and locations. Only sensors that recorded all MS events were retained for analysis. Specifically, if the total number of MS events is denoted as N_e, any sensor that has recorded fewer than N_e events is excluded. The number of sensors meeting this criterion is denoted as N_{sensor}.

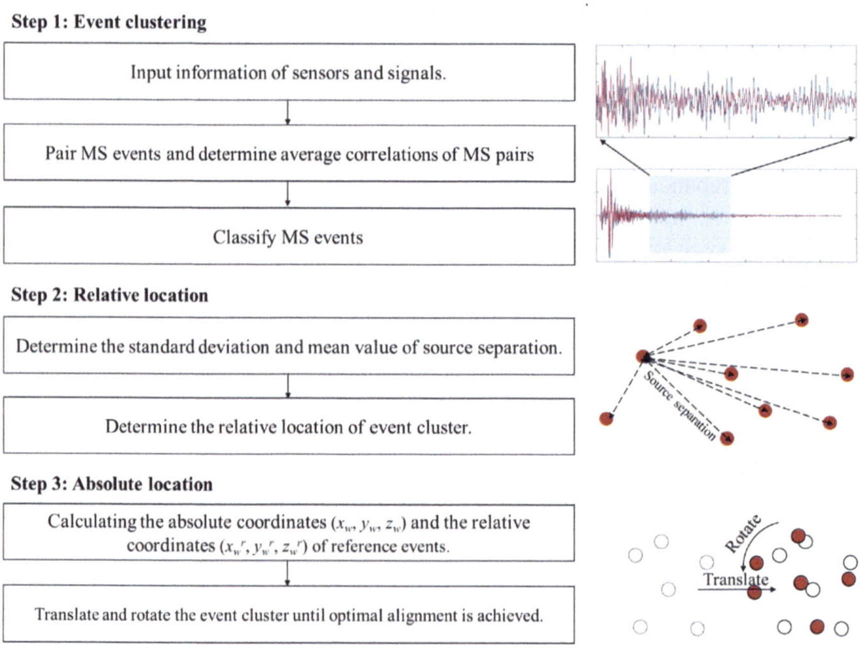

Fig. 5.5 Flowchart of the CWI-based MS location method in a tunnel

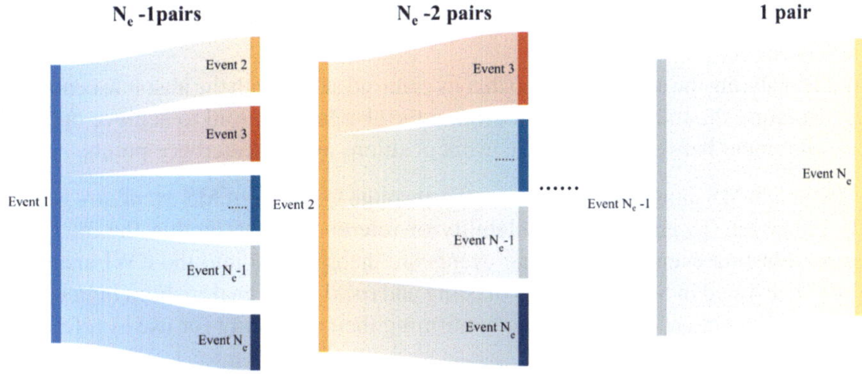

Fig. 5.6 Mode of pairing MS events

All MS events are paired according to the scheme illustrated in Fig. 5.6. From N_e events, a total of $(N_e^2 - N_e)/2$ unique event pairs can be generated. Each event pair comprises N_{sensor} signal pairs, corresponding to the sensors common to both events. For each signal pair, the cross-correlation is calculated using Eq. 5.3, and the average of these values is taken as the correlation coefficient for the event pair. The time window employed in this analysis follows the settings described in Sect. 5.2.3.

All event pairs are then sorted in descending order based on their correlation values. Those with a correlation below a predefined threshold C_r are discarded. An event pair is considered to represent two correlated events. MS events are subsequently clustered according to the following rule: if event a is correlated with event b and event b is correlated with event c, then events a, b, and c are regarded as belonging to the same cluster.

(2) **Relative Location**

Taking the point 150 ms after the P-wave arrival as the onset, a 200 ms window is defined. The wave in this window is used to determine the source separation by Eqs. 5.4–5.9. The source separation determined from each sensor is incorporated into Eq. 5.10. When the predicted locations achieve optimal alignment with the source separation, the output coordinates of all events are identified as the relative locations of the event clusters.

(3) **Absolute Location**

When an event cluster contains reference events with known absolute locations, the relative coordinates of the cluster can be transformed into absolute positions. In two-dimensional space, at least two events with known locations are needed; in three-dimensional space, a minimum of three such events is necessary.

Let the known absolute locations be denoted as (x_i^0, y_i^0, z_i^0), where $i = 1, 2 \ldots, N_e^0$, and N_e^0 is the number of events with known coordinates. The transformation procedure involves three main steps:

(a) Computing the centroid of the absolute coordinates and that of the relative coordinates.
(b) Translating the event cluster so that its centroid aligns with the absolute centroid;
(c) Rotating the translated cluster around the absolute centroid to achieve optimal alignment between the relative event positions and the reference points.

In typical MS monitoring scenarios, the absolute locations of MS events are generally unknown. To ensure the availability of reference events within the analyzed cluster, blasting events or rockburst events are incorporated into the CWI analysis. As demonstrated in Sects. 5.2, both blasting and rockburst signals exhibit high correlation coefficients with MS signals, confirming their suitability for use as reference events.

5.4 Numerical Validations

5.4.1 Preparation for Numerical Testing

The location performance of the CWI method is assessed through numerical simulations. A synthetic model domain of dimensions 50 m × 100 m is established with a grid resolution of 0.1 m × 0.1 m. The viscoelastic wave equations are solved numerically via the finite-difference method. Detailed descriptions of the equations and their numerical implementation are provided in Bohlen (2022). To simulate wave propagation in an unbounded medium while preserving computational efficiency, a PML absorbing boundary condition (Collino and Tsogka 2001) is applied at the model boundaries to suppress spurious reflections. In this study, the domain frequency is set to 200 Hz. For more detailed simulation parameters of the synthetic test site, please refer to Table 5.2. For the purpose of evaluating the performance of the CWI method, the P-wave velocity in this section is 5000 m/s, which facilitates calculation.

In the numerical model illustrated in Fig. 5.7, both MS sources and sensors are positioned on the tunnel's same side to guarantee reliable detection of stress wave signals while minimizing significant diffraction effects. Furthermore, forty rock discontinuities featuring varying lengths, widths, and orientations are randomly placed throughout the model to realistically capture the heterogeneous nature of tunnel environments.

Table 5.2 Parameters of the numerical simulation

Medium	Velocity (m/s)		Density(kg/m^3)	Quality factor	
	P-wave	S-wave		P-wave	S-wave
Rock	5000	3000	2500	30	30
Tunnel (air)	340	0	7.9	20	20
Rock discontinuities	340	0	7.9	20	20

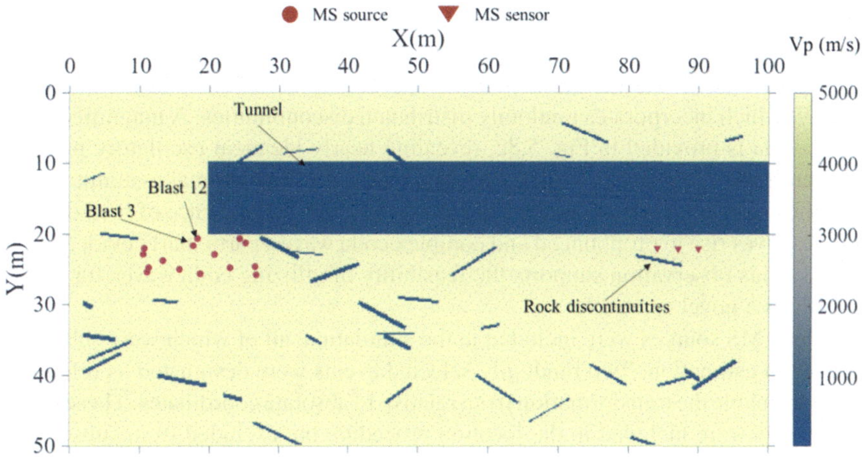

Fig. 5.7 The model with forty rock discontinuities for numerical testing. Detailed source and sensor coordinates are provided in Table 5.3

Table 5.3 Coordinates of sources and sensors

Type	*x* (m)	*y* (m)	Type	*x* (m)	*y* (m)
Source 1	24.2	20.6	Source 10	13.2	23.8
Source 2	20.6	22.8	Source 11	11.2	24.8
Source 3	17.6	21.6	Source 12	18.2	20.6
Source 4	15.6	25	Source 13	22.8	24.8
Source 5	25	21.2	Source 14	18.6	25.6
Source 6	26.4	24.6	Source 15	10.4	22.8
Source 7	10.6	22	Sensor 1	84	22
Source 8	26.6	25.8	Sensor 2	87	22
Source 9	11	25.4	Sensor 3	90	22

Fifteen MS sources are randomly distributed near the working face, as shown in Fig. 5.7. Three sensors are installed approximately 60 m behind the working face, with a spacing of 3 m between adjacent sensors. The coordinates of the MS sources and sensors are shown in Table 5.3.

5.4.2 Numerical Test Results of the CWI Method

Figure 5.8b displays the waveforms recorded by sensors for sources 3 and 12 in a model that includes only the tunnel and a homogeneous rock medium (i.e., without any rock discontinuities). In this case, the P- and S-wave arrivals are clearly distinguishable, and the waveform amplitude decays significantly after approximately 80 ms. Figure 5.8a shows the corresponding waveforms obtained from the model in Fig. 5.7, which incorporates randomly distributed discontinuities. A magnified view of Fig. 5.8a is provided in Fig. 5.8c, revealing nearly identical oscillatory patterns between the two signals, a finding consistent with the fundamental assumptions of CWI theory. A comparison of Fig. 5.8a, b indicates that the presence of rock discontinuities gives rise to pronounced and complex coda waves, particularly evident after 100 ms. This observation supports the feasibility of utilizing coda waves for signal analysis as a novel approach.

Fifteen MS sources were included in the simulation, all of which were subjected to location estimation. Two randomly selected events were designated as reference events to enable the transformation from relative to absolute coordinates. These reference events were included in the location procedure but excluded from subsequent error analysis. Since the primary objective of this numerical test was to evaluate the feasibility of the CWI method, the P-wave velocity required for CWI analysis was directly adopted from Table 5.2 (5000 m/s). Additionally, owing to the shorter signal duration inherent in numerical simulations compared to field conditions, the coda window used for CWI analysis was limited to 100 ms.

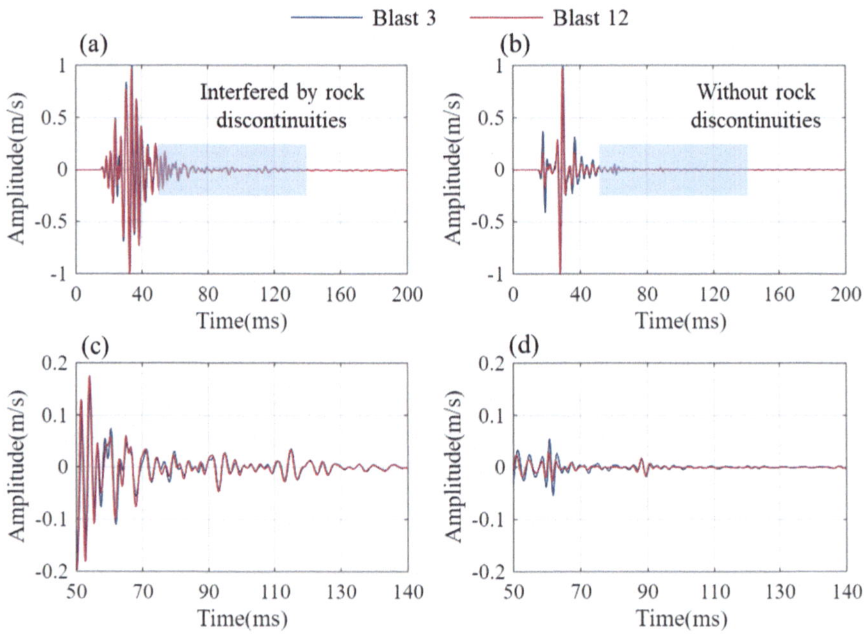

Fig. 5.8 The waveforms of MS sources 3 and 12. **a** Signals generated from a model with rock discontinuities. **b** Signals generated from the model without rock discontinuities. **c** Signals between 50 and 140 ms in (**a**). **d** Signals between 50 and 140 ms in (**b**)

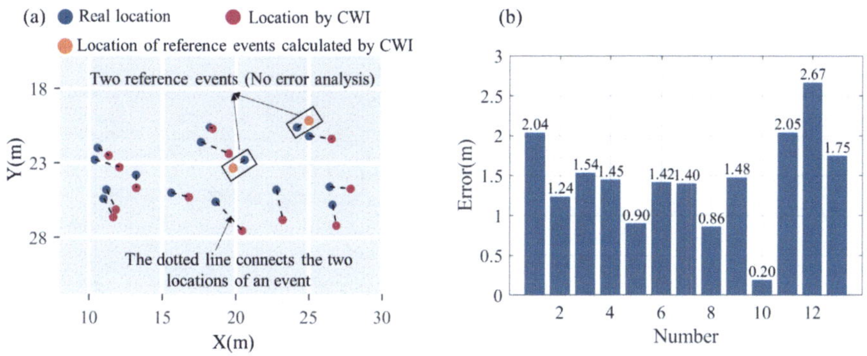

Fig. 5.9 The location results of 13 sources, excluding reference events. **a** Coordinates of the real location and calculated location. **b** Error analysis

The resulting source locations are presented in Fig. 5.9a, where blue and red markers indicate the true positions and CWI-derived locations, respectively. Dashed lines connect the corresponding true and estimated coordinates for visual comparison. The two reference events, marked by black rectangles in Fig. 5.9a, are excluded from error quantification. Location errors, defined as the Euclidean distance between true and estimated coordinates, are summarized in Fig. 5.9b. For the remaining 13 events, the maximum, minimum, and average location errors were 2.67 m, 0.20 m, and 1.46 m, respectively. These results demonstrate that the CWI method achieves high precision and stability in locating MS event clusters. The simulation outcomes preliminarily validate the applicability of the CWI method for tunnel-based MS event cluster localization.

5.5 Field Study

5.5.1 The Velocity Determination Applicable to the CWI Method

Given the difficulty of directly measuring the P-wave propagation velocity in the surrounding rock, this study estimates it using blasting event signals. However, the simple ratio of source-to-sensor distance to P-wave travel time is unsuitable, as the CWI method requires intersource velocity. Therefore, an alternative approach is adopted for a reliable approximation.

The MS monitoring system in the Bayu tunnel consists of two arrays with eight sensors in total. During deployment, the relative positions of sensors in each array are fixed. As shown in Fig. 5.10a, sensors S1 - 1 and S2 - 1 have the same relative orientations to the tunnel alignment, with nearly the same y and z coordinates but different x-axis displacements. In this setup, four sets of corresponding sensors, one

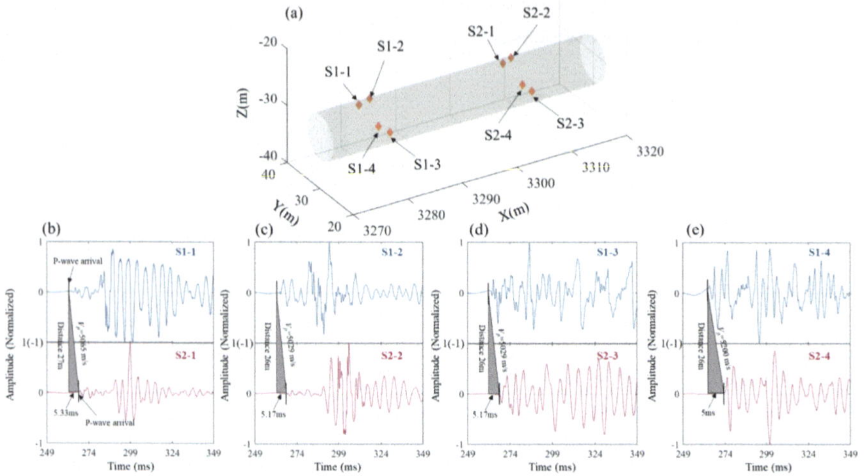

Fig. 5.10 Determination of P-wave velocity applicable to the CWI method. **a** The layout of eight sensors in the Bayu tunnel. **b** Signals recorded by S1-1 and S2-1. **c** Signals recorded by S1-2 and S2-2. **d** Signals recorded by S1-3 and S2-3. **e** Signals recorded by S1-4 and S2-4

from each array, are paired according to their spatial relationships. Then, as shown in Fig. 5.10b–e, the P-wave velocity is estimated using the distance between paired sensors and the corresponding difference in P-wave arrival times.

5.5.2 An Example of CWI Analysis

This section presents a case study illustrating the application of the CWI-based location method using the first 10 MS events recorded on May 8, 2019. Blasting events that occurred between May 8 and May 10 are incorporated as reference events, bringing the total number of events analyzed to 13.

The waveforms of these 13 events were recorded by six sensors. All events were paired according to the scheme shown in Fig. 5.6, resulting in a total of 78 event pairs. For each event pair, the signals recorded by a common sensor formed six signal pairs. The average correlation coefficient across these signal pairs was taken as a measure of similarity between the two events; the results are presented in Fig. 5.11. Correlation coefficients exceeding 0.5, 0.7, and 0.9 were observed in 89.74%, 25.64%, and 1.28% of the signal pairs, respectively. In this analysis, all signal pairs with correlation coefficients below 0.5 were discarded. Subsequent evaluation of the remaining pairs led to the classification of all 13 events into a single cluster.

The source separations estimated from each sensor were integrated into Eq. 5.12 to establish the spatial relationships among the MS events. The predicted source

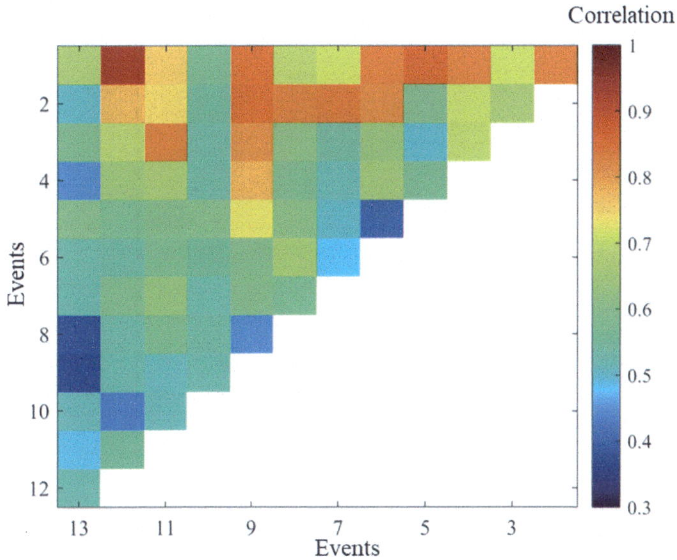

Fig. 5.11 Cross-correlation analysis of 78 event pairs generated by 13 events

locations were iteratively adjusted until optimal consistency with the estimated separations was achieved. The resulting coordinate solutions define the relative spatial distribution of the events within the cluster. Figure 5.12 presents the location results for the event cluster, illustrating both the relative locations and the subsequent transformation to absolute locations. The transformation involved two sequential steps: a translation of $\Delta x = 3.72$ m, $\Delta y = -4.32$ m, and $\Delta z = 3.87$ m along the respective axes (Fig. 5.12e–h), followed by a rotation of $\theta x = 66.70^{\circ}$, $\theta y = 321.09^{\circ}$, and $\theta z = 45.71^{\circ}$ about each axis (Fig. 5.12i–l). The orange points in Fig. 5.12 indicate the recovered locations of the reference events obtained via the CWI method. The relative spatial configuration of the three blasting sources is accurately reconstructed, maintaining good agreement with their actual positions.

5.5.3 Assessing the Accuracy of the CWI Method

This section applies the CWI method to locate all MS events recorded in the Bayu tunnel from May 8 to May 18, 2019. As the CWI method is specifically designed for locating clusters of MS events, the events are grouped into sets of 15 for each location run, with three blasting events included in each set to serve as reference events. Two scenarios are designed to assess the performance of the CWI method: (1) the location of blasting and rockburst events and (2) the location of MS events.

The source coordinates of blasting and rockburst events are determined based on the center of the corresponding burst crater. This practice of using crater centers as

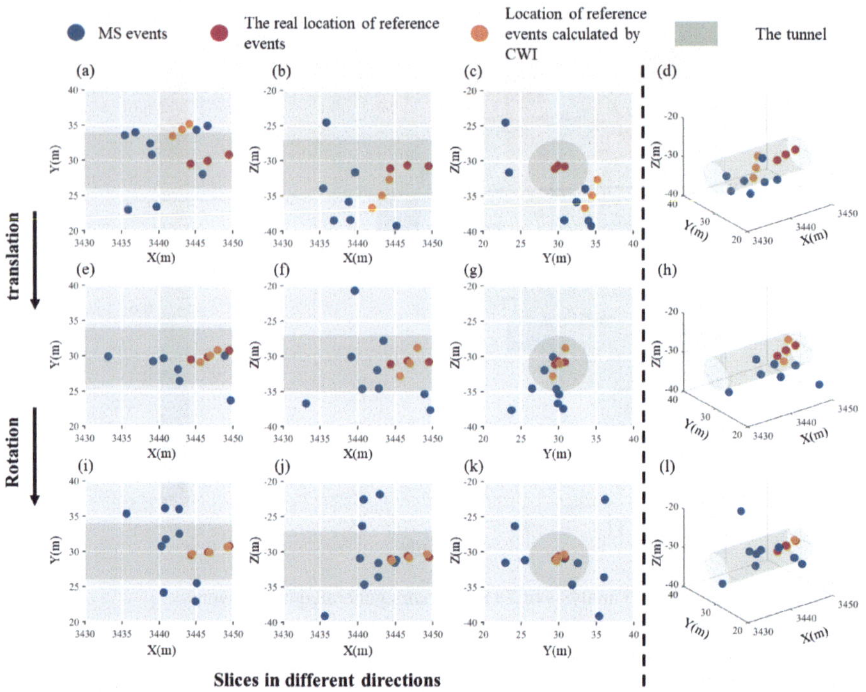

Fig. 5.12 The process of converting the absolute locations of 13 events. **a** XOY slice, **b** XOZ slice, **c** YOZ slice, and **e** 3D view of the relative location determined by CWI. **e** XOY slice, **f** XOZ slice, **g** YOZ slice, and **h** 3D view of translated locations. **i** XOY slice, **j** XOZ slice, **k** YOZ slice, and **l** 3D view of absolute locations

reference source locations has been widely adopted in previous studies to validate arrival-time picking algorithms and assess location accuracy (Feng et al. 2015; Zhang et al. 2021). Accordingly, the present study adopts this approach for verification purposes.

Scenario 1: Blasting and Rockburst Event Location

Although Scenario 1 focuses on evaluating the location results of blasting and rockburst events, the computational procedure does not rely solely on these signals. The CWI method is consistently applied to groups of 15 events, with only the blasting and rockburst events subsequently extracted for error analysis. In principle, location using only four events, three references and one unknown, is theoretically possible. However, such configurations tend to yield unreliable results. A detailed discussion of this limitation is provided later in the study.

A comparison of the CWI-derived locations with the crater centers of eight blasting events and four rockburst events is presented in Fig. 5.13. The location errors range from 1.00 m to 4.71 m for blasting events and from 2.71 m to 4.57 m for rockburst events. The corresponding mean errors are 2.95 m and 3.65 m, respectively, indicating relatively higher accuracy for blasting events. Overall, the CWI method

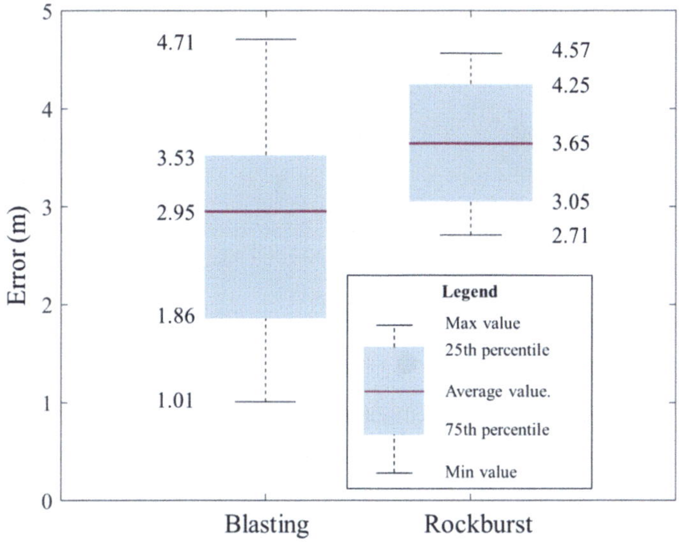

Fig. 5.13 Errors in the location of blasting and rockburst events

achieves a maximum error of less than 5 m and a mean error of approximately 3 m, confirming its localization accuracy and preliminary feasibility for drill-and-blast tunneling applications.

Scenario 2: Comprehensive MS Event Location

In this scenario, event clusters are located using both the CWI method and the IMS Trace software (Salvoni et al. 2016) for comparative analysis. The IMS Trace software incorporates a commercially established MS monitoring and analysis system, whose source location module relies on conventional traveltime difference methods based on a homogeneous velocity model. Figure 5.14a presents the locations of all MS events in the Bayu Tunnel obtained via the CWI method during the study period, while Fig. 5.14b displays the corresponding results from the IMS Trace software. Both approaches reveal high concentrations of MS events near the working face and rockburst zones.

To quantitatively assess the clustering characteristics, daily event centroids are calculated as the mean coordinates of MS events, and the distance from each event to its respective daily centroid is analyzed (Fig. 5.15). For the CWI-based results, only 1.39% of events lie more than 25 m from their centroids, a markedly lower proportion than the 9.42% observed with the IMS Trace software. Both the mean distances and the compactness of the clusters indicate that the CWI method achieves superior performance. This comparative analysis demonstrates that the CWI method yields more tightly concentrated event distributions and improved location quality. Together with the findings from Scenario 1, where location errors for rockbursts were within 4.57 m, these results confirm that the CWI method provides a reliable, stable, and accurate approach for locating MS event clusters.

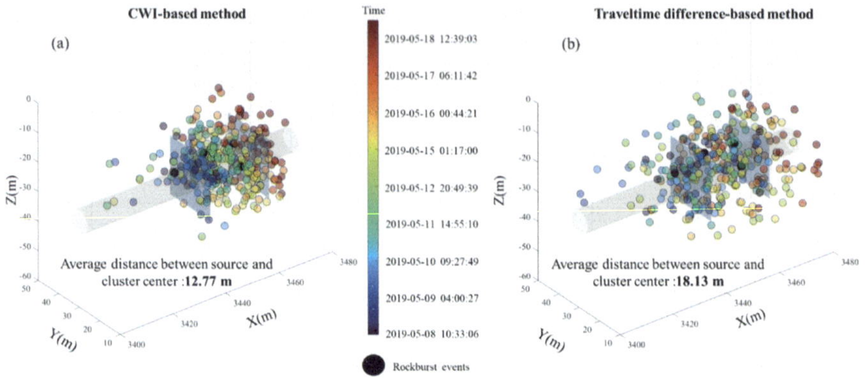

Fig. 5.14 The results of MS events located by the **a** CWI-based method and **b** traveltime difference-based method in the Bayu tunnel on May 8 and May 18, 2019

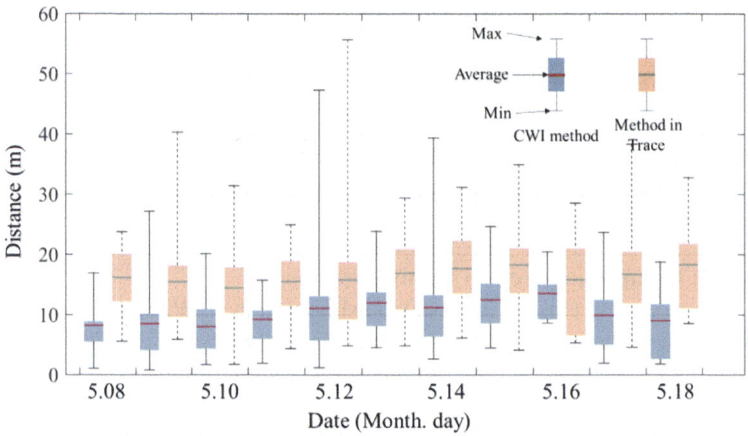

Fig. 5.15 Distance between events and daily cluster center (calculated as the mean coordinates of daily MS events)

5.6 Discussion

Numerical simulations and field applications demonstrate that the CWI method is a feasible approach for locating MS event clusters in tunnels, exhibiting both high accuracy and stability. Nevertheless, several operational considerations and methodological aspects warrant further investigation.

(1) **Coda Wave Calculation Length**

The CWI method requires an appropriately selected coda wave analysis window. In the field application described above, a fixed window length of 200 ms was adopted. To evaluate the influence of this parameter, location analyses were performed using

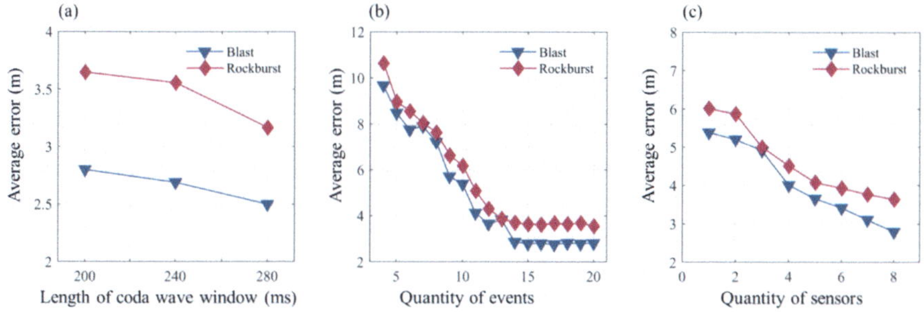

Fig. 5.16 Parameter sensitivity analysis. **a** Length of coda wave, **b** quantity of events, and **c** quantity of sensors

coda lengths of 200 ms, 240 ms, and 280 ms on both blasting and rockburst signals. As shown in Fig. 5.16a, the 280 ms window yielded the smallest deviations, 2.5 m for the blasting event and 3.17 m for the rockburst event, indicating that longer coda windows contribute to improved location accuracy.

However, it remains unclear whether the coda wavelength can be increased indefinitely to further reduce location errors. Given the limited duration of coda waves in MS signals, excessively long analysis windows may incorporate invalid waveform segments, potentially degrading performance. Thus, future research could focus on developing adaptive coda length determination algorithms capable of dynamically optimizing the analysis window for each event group.

(2) **Number of Events in Each Location**

The CWI method requires a minimum of four events for location, three serving as references and one as the target. To assess the influence of the number of events included, the "5.08" rockburst event and a blasting event were analyzed with varying quantities of additional MS events, while three blasting events (occurring between May 9 and 11) were consistently used as known references.

As illustrated in Fig. 5.16b, when the total number of events exceeded 14, the location errors for both the blasting and rockburst events remained below 4 m. In contrast, analyses conducted with only the three reference events and a single target event resulted in poor performance, with errors approaching approximately 10 m. These findings indicate that the effectiveness of the CWI method improves substantially with larger event groups.

(3) **Impact of Sensor Quantity**

To examine the dependence of the CWI method on the number of available sensors, the locations of the "5.08" rockburst event and the blasting event on May 8 were recalculated using sensor counts ranging from one to eight. As illustrated in Fig. 5.16c, the location errors for both event types decrease notably as the number of sensors increases from one to eight. Although the CWI method remains functional even with a single sensor, its accuracy improves substantially with additional sensors. This characteristic renders the method particularly suitable for sparse-array monitoring configurations but entails a necessary trade-off between deployment simplicity and localization precision.

(4) **Guidance for Parameter Determination**

We examined the sensitivity of three parameters: coda wave length, number of events, and number of sensors. To achieve optimal performance with the CWI method, the following guidelines should be observed:

a. Coda wave calculation length: This requires employing a relatively long window while ensuring that it remains shorter than the effective length of the signal.
b. Quantity of events: This value needs to be larger than 15. A small number of MS events will reduce the accuracy.
c. Quantity of sensors: Field implementation requires maximizing the number of sensors.

(5) **Hybrid Use of Blasting and Rockburst Events**

The analyses presented in Figs. 5.14 and 5.16 consistently reveal that rockburst events yield higher location errors than blasting events. This finding appears counterintuitive, given that Sect. 5.2.4 demonstrates stronger waveform correlation between rockbursts and MS signals. The discrepancy can be attributed to the exclusive use of blasting events as known reference events in the Bayu tunnel case. To enhance the long-term performance of the CWI method in tunnel environments, this study recommends incorporating rockburst events or hybrid datasets combining rockburst and blasting events as reference sources. This strategy exploits the inherent correlation between rockburst signals and MS activity, thereby potentially improving the overall location accuracy.

5.7 Conclusion

This chapter presents an absolute location method for MS event clusters based on coda wave interferometry. The proposed approach exhibits low dependence on the velocity model and enables the relative location of MS event clusters even with a sparse sensor array. By incorporating reference events into the interferometric analysis, the relative locations are subsequently transformed into absolute coordinates. The methodology comprises three main steps: event clustering, relative location, and absolute location. Its feasibility and accuracy have been validated through numerical simulations and field application in the Bayu Tunnel, where its performance was compared with commercial IMS software. The main conclusions are summarized as follows.

Analysis of MS monitoring data from the Bayu Tunnel revealed two key characteristics: the distribution of coda wavelengths and the correlation between MS signals and those from blasting or rockburst events. Blasting signals predominantly exhibit coda lengths between 400 and 700 ms, while rockburst signals show shorter coda waves ranging from 200 to 400 ms. MS signals display a more uniform distribution of coda lengths. Significant correlations were identified between MS signals and both blasting and rockburst signals, with a particularly strong correlation observed between MS and rockburst signals. These findings confirm the feasibility of applying coda wave interferometry jointly to blasting, rockburst, and MS signals and suggest an optimal coda window duration of 200 ms.

A two-dimensional tunnel model incorporating multiple rock discontinuities was developed to test the CWI method. The resulting maximum, minimum, and average location errors were 2.67 m, 0.20 m, and 1.46 m, respectively, effectively demonstrating the feasibility of the proposed method for tunnel applications.

The method was applied to the Bayu Tunnel under two test scenarios. In Scenario 1, the location errors for blasting and rockburst events averaged 2.95 m and 3.65 m, respectively. In Scenario 2, MS event locations obtained via the CWI method were compared with those from commercial MS software based on a conventional homogeneous velocity model and traveltime difference location. The comparative analysis confirmed that the CWI-based approach yields improved location accuracy relative to traditional methods, fully validating its enhanced precision and stability.

Using data from the Bayu tunnel, the influence of key parameters was investigated, including the coda wave window length, number of events within a cluster, and number of deployed sensors. The results indicate that extending the coda window may improve location accuracy. Furthermore, including more than 14 events per cluster can reduce errors to approximately 3.5 m. Although the method remains applicable with sparse sensor arrays, even with a single sensor, such configurations may compromise accuracy. Better performance is achieved with at least three sensors, maintaining location deviations at approximately 5 m.

References

Barajas A, Margerin L, Campillo M (2022) Coupled body and surface wave sensitivity kernels for coda-wave interferometry in a three-dimensional scalar scattering medium. Geophys J Int 230(2):1013–1029

Bohlen T (2002) Parallel 3-D viscoelastic finite difference seismic modelling. Comput Geosci 28(8):887–899

Chen B, Feng X, Li S, Yuan J, Xu S (2009) Microseism source location with hierarchical strategy based on particle swarm optimization. Chin J Rock Mechan Eng 28(4):740–749

Chen BR, Wu H, Chi XW, Liu H, Wu MD, Yan JW (2019) Real-time recognition algorithm for microseismic signals of rock failure based on STA/LTA and its engineering application. Rock Soil Mech 40(9):3689–3696

Chen Y, Savvaidis A, Fomel S, Saad OM, Chen Y (2023) RFloc3D: a machine-learning method for 3-D microseismic source location using P-and S-wave arrivals. IEEE Trans Geosci Remote Sens 61:1–10

Collino F, Tsogka C (2001) Application of the perfectly matched absorbing layer model to the linear elastodynamic problem in anisotropic heterogeneous media. Geophysics 66(1):294–307

Ding L, Gao EG, Liu Q, Qian W, Liu D, Yang C (2018) Reverse-time ray-tracing method for microseismic source localization. Geophys J Int 214(3):2053–2072

Fard RA, Javan-Doloei G, Farrokhi M, Rahimi H, Mahood M (2020) Coda wave attenuation's dependency on Lapse time and frequency in west of Iran plateau using local earthquakes. Ann Geophys 63(4):SE437–SE437

Feng GL, Feng XT, Chen BR, Xiao YX, Jiang Q (2015) Sectional velocity model for microseismic source location in tunnels. Tunnell Underground Space Technol 45:73–83

Geiger L (1912) Probability method for the determination of earthquake epicentres from the arrival time only. Bull St. Louis Univ 8:60

He M, Canbulat I, Suorineni F et al (2025) Definition and classification of rockburst. Rock Mech Bull 100206

Jiang R, Dai F, Liu Y, Li A (2021) Fast marching method for microseismic source location in cavern-containing rockmass: performance analysis and engineering application. Engineering 7(7):1023–1034

Li B, Xu N, Xiao P, Xia Y, Zhou X, Gu G, Yang X (2023) Microseismic monitoring and forecasting of dynamic disasters in underground hydropower projects in southwest China: a review. J Rock Mech Geotech Eng 15(8):2158–2177

Li S, Kuang Z, Xiao Y, Qiao Z, Yang W (2022) Rockburst tendency prediction based on an integrating method of combination weighting and matter-element extension theory: a case study in the Bayu Tunnel of the Sichuan–Tibet Railway. Eng Geol 308:106796

Liu L, Li S et al (2024) Identification of rock discontinuities by coda wave analysis while borehole drilling in deep buried tunnels. Tunnell Underground Space Technology 153:105969

Peng P, Wang L (2019) Targeted location of microseismic events based on a 3D heterogeneous velocity model in underground mining. PLoS One 14(2):e0212881

Robinson DJ, Sambridge M, Snieder R (2011) A probabilistic approach for estimating the separation between a pair of earthquakes directly from their coda waves. J Geophys Res: Solid Earth 116(B4)

Salvoni M, Morkel IG, Dight PM 2016. Microseismic data—A comparison between routine trigger method and continuous data processing. In: APSSIM 2016: proceedings of the First Asia Pacific slope stability in mining conference, pp 639–655.

Snieder R (2006) The theory of coda wave interferometry. Pure Appl Geophys 163(2):455–473

Snieder R, Vrijlandt M (2005) Constraining the source separation with coda wave interferometry: Theory and application to earthquake doublets in the Hayward fault, California. J Geophys Res: Solid Earth 110(B4)

Snieder R, Grêt A, Douma H, Scales J (2002) Coda wave interferometry for estimating nonlinear behavior in seismic velocity. Science 295(5563):2253–2255

Waldhauser F (2009) Near-real-time double-difference event location using long-term seismic archives, with application to Northern California. Bull Seismol Soc Am 99(5):2736–2748

Waldhauser F, Ellsworth WL (2000) A double-difference earthquake location algorithm: method and application to the northern Hayward fault, California. Bull Seismol Soc Am 90(6):1353–1368

Wamriew D, Charara M, Pissarenko D (2022) Joint event location and velocity model update in real-time for downhole microseismic monitoring: A deep learning approach. Comput Geosci 158:104965

Wang J, Chen G, Xiao Y, Li S, Chen Y, Qiao Z (2021) Effect of structural planes on rockburst distribution: case study of a deep tunnel in Southwest China. Engineering Geology 292:106250

Xiao Y, Chen S, Liu L et al (2024) Determining S-wave arrival times for microseismic monitoring in tunnels based on instantaneous phase difference of sensors. Rock Mech Rock Eng 57(3):1781–1797

Xiao Y, Chen S, Feng G et al (2025) Picking P-wave arrival times from microseismic signals in tunnels with low signal-to-noise ratios without noise filtering. Soil Dyn Earthquake Eng 192:109288

Xiao Y, Chen S, Liu L et al (2026) A stratified double-difference dynamic tomography of 3D velocity field for deep buried large underground cavern. Rock Mech Rock Eng. https://doi.org/10.1007/s00603-025-05213-0

Zhang H, Thurber CH (2003) Double-difference tomography: the method and its application to the Hayward fault, California. Bull Seismol Soc Am 93(5):1875–1889

Zhang J, Xu N, Wu W, Xiao P, Li B, Dong L (2022) Improvement of microseismic source location during cavern excavation in faulted rock mass using fast marching method. Tunnell Underground Space Technol 128:104661

Zhang P, Xu N, Xia P, Zhao T, Gao F, Ding X, Li B (2023) Microseismic source location based on improved artificial bee colony algorithm: performance analysis and case study. J Intell Constr 1(3):1–15

Zhang W, Feng XT et al (2021) An arrival time picker for microseismic rock fracturing waveforms and its quality control for automatic localization in tunnels. Comput Geotech 135:104175

Zhao Y, Curtis A (2019) Relative source location using coda-wave interferometry: method, code package, and application to mining-induced earthquakes. Geophysics 84(3):F73–F84

Chapter 6
MS Source Location in Deep-Buried Tunnels Using the Updated Inverted Heterogeneous Velocity

6.1 Prologue

In Chap. 5, we proposed the CWI method. This method can perform MS event group location under sparse sensor array conditions. However, the real-time performance of the CWI method is poor, especially when there are few MS events. In this case, the traditional time difference location method still needs to be used.

It is well known that the accuracy of time difference location methods depends on the surveyed region's velocity model. A uniform-velocity model may work for simple lithology projects but causes location inaccuracies in complex geological settings (Xiao et al. 2026; Liu et al. 2023). To solve these issues, several approaches have been developed. Feng et al. (2015) presented a segmented velocity model for MS localization and applied it in multiple tunnels. Castellanos and Van der Baan (2015) introduced a cross-correlation-based location method for mining. Collins et al. (2014) developed a velocity model considering different void infill materials. Dong et al. (2011, 2017a, b, 2020) developed an MS location method that reduced the effect of velocity uncertainty, complementing traditional location theory. However, MS source locations in deep tunnels still face challenges. First, complex velocity structures such as the excavation damaged zone (EDZ) change seismic wave propagation paths, so the velocity model needs to capture EDZ characteristics for accurate traveltime calculation. Second, the acquisition geometry of tunnel MS monitoring networks is ill posed, and common velocity analysis and inversion methods are not applicable to tunnel linear sensor layouts. Third, the surrounding rock condition changes rapidly during excavation, requiring continuous updates to the velocity model for location accuracy.

Blasting sources in tunnel excavation provide valuable waveform data for active seismic surveying without precise trigger times. Experimental and numerical studies show that the EDZ under a concrete lining can be characterized by surface waves along tunnel sidewalls. The S-wave velocity model can be derived by inverting the fundamental mode of Rayleigh waves, and then the blasting trigger times can be

© Tongji University Press Co., Ltd 2026

Y. Xiao et al., *Near-Field Microseismic Data Processing in Deep Tunnels*,
https://doi.org/10.1007/978-981-92-3208-6_6

estimated from the inverted velocities. With multiple blasting sources, P-wave first-arrival traveltime tomography is possible for continuous inversion of heterogeneous P-wave velocity fields during tunnel excavation. However, it is challenging under ill-posed acquisition geometries and with few blasting sources. First-arrival traveltime tomography methods can be divided into traditional ray tracing-based and adjoint-state methods, depending on explicit or implicit ray path calculations. Ray tracing-based methods use forward modeling operators such as straight-ray, curved-ray, or shortest-path ray tracing. Inversion is usually performed by the least-squares conjugate-gradient algorithm or simultaneous iterative reconstruction, with velocity model updates from solving tomographic equations. Although the theoretical framework of ray tracing-based inversion is well established, the tomographic equations are often ill-conditioned, causing solution instability.

Taillandier et al. (2009) presented the adjoint-state method for first-arrival traveltime tomography, which integrates implicit ray tracing, specifically for surface seismic surveys. Within this method, either the fast marching method (FMM) or the fast sweeping method (Zhao 2005) is utilized to solve the eikonal equation, and then the gradient is ascertained through dual modeling. When compared to techniques rooted in ray tracing, the adjoint-state method provides an improved capacity for imaging complex geological structures, especially when there are only a small number of source–receiver pairs accessible.

This study fully utilizes drilling and blasting source data to devise an MS source location method, integrating an updated inverted heterogeneous velocity model. We propose an enhanced perturbational dispersion inversion technique, merging a modified thin-layer method for forward simulation with a perturbational inversion framework. This method allows for the estimation of the S-wave velocity model and the EDZ depth. The initiation time of blasting sources is then ascertained via S-wave back-propagation tracing using FMM on the inverted S-wave model. An adjoint-state first-arrival tomography (AST) method is used to continuously invert the heterogeneous P-wave velocity field. A double-difference simplex-based method is utilized for MS source localization. To verify the proposed approach, a realistic numerical simulation based on the finite-difference time-domain method is performed to visualize the blasting source wavefield in the surrounding rock. The location method's accuracy is examined using 64 MS sources distributed around the tunnel working face. Moreover, a field experiment is conducted at two newly excavated tunnels with a burial depth of 2400 m at the Jinping Underground Laboratory. Blasting events with known source locations are employed to assess the practical performance of the proposed location method under real-world conditions.

6.2 Methodology

6.2.1 Motivation

Obtaining a reliable velocity model of the monitored area is a fundamental prerequisite for accurate MS source location. The velocity model governs wave propagation paths and arrival times, directly influencing localization accuracy. The presence of the EDZ, together with rock mass discontinuities and variations in integrity and stress conditions, results in complex wave propagation behavior. Figure 6.1 illustrates the effect of the EDZ on wave propagation. Under a homogeneous velocity assumption, wave paths are represented as straight lines connecting sources and sensors (Fig. 6.1a). However, when the EDZ is considered, a distinct low-velocity zone appears adjacent to the existing tunnel, as shown in Fig. 6.1b. In this heterogeneous model, propagation paths computed using FMM follow Fermat's principle of least time. Waves emitted from the source travel toward high-velocity regions before bending back to the sensors, resulting in path lengths exceeding those of straight-line assumptions. Consequently, adopting a homogeneous velocity model can introduce significant errors in MS event location.

In the scenario depicted in Fig. 6.1, a new tunnel is being excavated adjacent to an existing tunnel constructed years earlier. The development of the EDZ and its influence on actual wave propagation paths cannot be overlooked in the MS location. Nevertheless, it is feasible to invert the velocity model using drilling and blasting sources from the new tunnel. Such sources enable active seismic exploration even without precise knowledge of the source initiation time. Dispersion analysis of surface waves propagating along the tunnel sidewall and first-arrival traveltime tomography offer two promising techniques for inverting S-wave and P-wave velocities, respectively. Critical steps in these inversions include extracting dispersion curves from limited sensor data and determining the source initiation time t_0 for tomography. This study develops an integrated approach combining velocity model

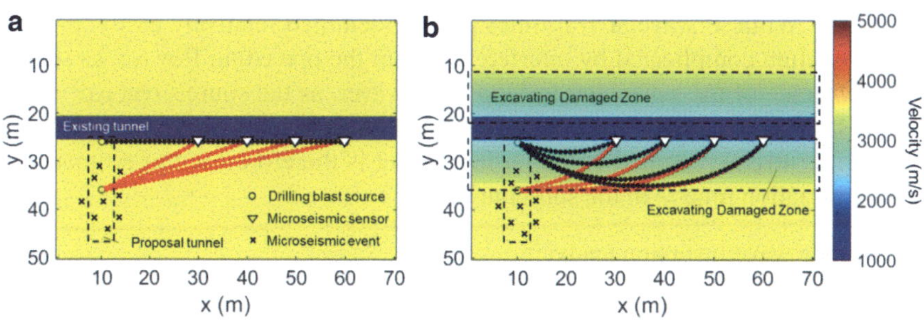

Fig. 6.1 Impact of the heterogeneous velocity on wave propagation (using the EDZ of the tunnel as an example): wave propagation routes in the **a** homogeneous velocity model and **b** heterogeneous velocity model

inversion with a double-difference MS location method. The proposed methodology aims to reconstruct the heterogeneous velocity field using ongoing drilling and blasting signals and to improve MS location accuracy by leveraging the updated inverted velocity model.

6.2.2 Overall Workflow

Figure 6.2 illustrates the proposed workflow for inverting heterogeneous velocity and source location, which comprises four main steps. First, the arrival times of P-waves and S-waves are obtained from the processed drilling-blasting signals. Dispersion inversion estimates a layered S-wave velocity model, which reflects the development of the EDZ and the geological stratification along the depth direction (y-axis in Fig. 6.1). Using the known source and sensor coordinates, the S-wave arrival times, and the layer S-wave velocity model, the blasting initiation time t_0 is determined through wave backward propagation analysis. Since the P- and S-waves share the same origin time, the absolute P-wave traveltime can subsequently be derived. Next, traveltime tomography is performed using multiple drilling and blasting sources to construct a heterogeneous P-wave velocity model for the region between the sources and sensors. As excavation progresses, additional sources become available, allowing the velocity model to be refined. Finally, double-difference location is carried out using the P-wave arrival times and the most recently updated heterogeneous P-wave velocity model. Each of the four steps is described in detail in the following sections.

(a) Drilling and blasting signal analysis

A bandpass filter is applied to suppress background noise and enhance the SNR. The drilling blast wave profile is illustrated in Fig. 6.2, where P-wave and S-wave arrivals can be distinguished based on their corresponding slopes. In practice, the drilling blast signal may extend over several seconds due to multiple blasting events. Only the first blast is analyzed in this study, as its first arrival is clean and free from overlapping waveforms. While P-wave arrival times can be identified relatively easily, S-wave picking is often complicated by interference from the preceding P-wave, as shown in the left trace of the schematic in Step 1. However, as the source–receiver offset increases, the separation between the P- and S-waves becomes more pronounced. The S-wave arrival recorded at the farthest offset is therefore used to estimate the blasting initiation time t_0 in the subsequent step.

(b) Surface wave inversion analysis

Surface wave dispersion inversion is employed to build the S-wave velocity model, enabling the calculation of t_0 through ray back-propagation, as illustrated in Step 2 of Fig. 6.2. The dispersion curve is derived from the preprocessed drilling and blasting wave profiles by the phase-shift method (Park et al. 1998). An initial model with linearly increasing S-wave velocity is assumed. By iteratively fitting the observed and

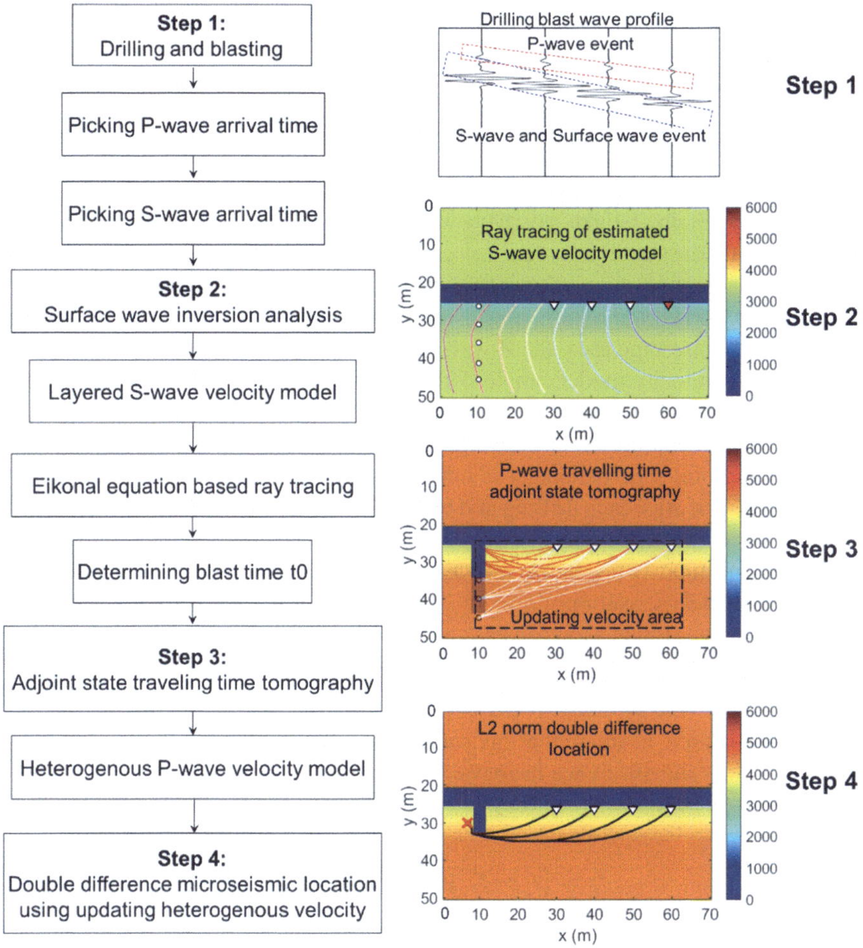

Fig. 6.2 Schematic of the proposed method for MS source location in deep tunnels incorporating velocity model updating

simulated dispersion curves, a one-dimensional S-wave velocity profile is obtained, which characterizes the extent of the EDZ. Notably, surface wave dispersion inversion does not require t_0. Once the S-wave velocity model is inverted, t_0 can be determined using the S-wave arrival time along with the known positions of the source and sensors. Ray tracing is performed on the inverted S-wave model by solving the eikonal equation via the FMM. The contour of traveltime from a given sensor (indicated by the red triangle in the Step 2 schematic) is shown in Fig. 6.2. The back-propagated traveltime derived from this analysis is then used to compute t_0 according to Eq. 6.1:

$$t_0 = t_s - t_b \tag{6.1}$$

where t_s is the S-wave arrival time at the farthest sensor and t_b is the back-propagated traveltime from that sensor to the source. This approach yields t_0 for blasting sources across different depths.

(c) Adjoint-state traveling time tomography (AST)

With the blasting initiation time t_0 determined, the absolute traveltime of the P-wave can be readily obtained. First-arrival traveltime tomography based on the AST is then employed to reconstruct the P-wave velocity model. As multiple blasting sources are incorporated into the tomographic inversion, the high-density ray paths progressively illuminate the monitored region, as illustrated in the schematic of Step 3 in Fig. 6.2. Within the ray coverage area, the P-wave velocity model is iteratively updated. An initial P-wave velocity model can be derived from the previously obtained S-wave velocity profile. The inversion proceeds by minimizing the misfit between observed and simulated traveltimes until convergence is achieved. To accelerate the iterative process, an adjoint-state parameter is introduced during the inversion. Tomography can be performed after each drilling and blasting cycle, allowing the P-wave velocity model to be refined incrementally as new blasting signals become available. In this manner, a continuously updated heterogeneous P-wave velocity model is maintained throughout tunnel excavation.

(d) Double-difference MS location using updating the heterogeneous velocity model

Source localization in tunneling environments relies on a heterogeneous P-wave velocity model. The difference in arrival times recorded by two sensors for the same event is defined as follows (Eq. 6.2):

$$r_{ij}^{k} = \left(t_i^k - t_j^k\right)^{\mathrm{obs}} - \left(t_i^k - t_j^k\right)^{\mathrm{cal}} \tag{6.2}$$

where r_{ij}^{k} denotes the arrival time difference between sensors i and j (where $i, j = 1,2 \ldots n$, and n is the total number of sensors) for MS event k. The terms Obs and Cal represent the observed and calculated arrival times, respectively. The FMM is employed to model traveltimes using an updated heterogeneous velocity model. Accordingly, the objective function for locating event k is formulated as the L2 norm of the arrival time differences:

$$F(x, y) = \sum_{i=1, j=2}^{n} \left(\left(t_i^k - t_j^k\right)^{\mathrm{obs}} - \left(t_i^k - t_j^k\right)^{\mathrm{cal}}\right)^2 \tag{6.3}$$

where (x, y) denotes the coordinates of the MS source. Minimization of the objective function (Eq. 6.3) is performed using the simplex method (Nelder and Mead 1965; Li et al. 2014), which iteratively adjusts and deforms the simplex shape to converge upon the location that minimizes the misfit.

The location method proposed in this study is designed to invert and update heterogeneous velocity models using seismic data from blasting events. Surface wave inversion is employed to account for the influence of the EDZ. The FMM is utilized to compute traveltimes and ray paths, as well as to estimate the blasting initiation time t_0. Traveltime tomography is subsequently performed to reconstruct a heterogeneous P-wave velocity model. As excavation progresses, additional blasting sources are incorporated into the tomographic inversion, enabling continuous model refinement. Finally, MS event locations are determined on the basis of a double-difference objective function combined with the simplex method. A detailed description of the surface wave inversion analysis, adjoint-state traveltime tomography, and double-difference MS location algorithm is provided below.

6.2.3 Surface Wave Inversion

The phase-shift method (Park et al. 1998) is employed to extract the dispersion curve from the blasting data. Subsequently, an improved perturbational inversion approach (Liu and Wang 2021) is introduced to invert a layered S-wave velocity model. This improved method comprises two main components: forward simulation using a modified thin-layer method (TLM) and the perturbational inversion procedure itself.

The modified TLM forward simulation first builds a matric function (Kausel and Roësset 1981):

$$K = \boldsymbol{k}^2\boldsymbol{B}_2 + k\boldsymbol{B}_1 + \boldsymbol{B}_0 - \boldsymbol{\omega}^2 M \tag{6.4}$$

In this formulation, k represents the wavenumber and $\boldsymbol{\omega}$ the angular excitation frequency. The matrices $\boldsymbol{B}_0$, $\boldsymbol{B}_1$, and $\boldsymbol{B}_2$ depend on Lamé's parameter and the shear modulus, whereas M is a function of density only. The equation $\mathbf{Det}(\boldsymbol{K}) = 0$ thus reduces to a generalized quadratic eigenvalue problem in k:

$$\left(k^2B_2 + kB_1 + B_0\right)v = \omega^2 M v \tag{6.5}$$

The equation is solved using a root-searching method to obtain the eigenvalue k and the corresponding eigenvector v, which together constitute an eigenvalue. For a given frequency ω, multiple eigenpairs may be obtained. The eigenvector v represents a mode shape, while the eigenvalue k, together with ω, defines the Rayleigh wave phase velocity as $V_\mathrm{R} = \frac{\omega}{k}$. In the modified thin-layer method, inverse iteration (Moler and Stewart 1973) is employed to compute the eigenvector v_j associated with each eigenpair (ω_j, V_{Rj}) for $j = 1, 2, \ldots, m$, thereby completing the forward simulation.

The perturbational inversion method subsequently updates the model parameters $\boldsymbol{m} = (\boldsymbol{V}_\mathrm{s}, \boldsymbol{V}_\mathrm{p}, \boldsymbol{\rho})$ by solving an optimization problem.

$$J(m)\Delta m = \Delta b \tag{6.6}$$

$\boldsymbol{\Delta b}$ is a function of frequency $\boldsymbol{\omega}$ and quantifies the misfit between observed and simulated dispersion curves. The Jacobian matrix $\boldsymbol{J(m)}$ is evaluated following Haney and Tsai (2017):

$$J(m) = \frac{\delta V_R}{\delta V_s}. \tag{6.7}$$

A weighting matrix W is defined for the optimization, formulated in terms of the absolute variation in Rayleigh wave velocity with frequency:

$$W = L^T L, L = \text{Diag}\left(\frac{|\Delta b|}{\|\Delta b\|_{\text{inf}}}\right). \tag{6.8}$$

In this expression, $|\Delta b|$ is the absolute value of the residual vector Δb, and $\|| \Delta b| \|_{\text{inf}}$ denotes its largest element. The diagonal matrix L is defined by the ratio $\frac{|\Delta b|}{\||\Delta b|\|_{\text{inf}}}$ along its diagonal. Incorporating this weighting matrix into Eq. 6.5 leads to:

$$A(m)\Delta m = d \tag{6.9}$$

where $A(m) = L^*J(m)$ and $d = L^*\Delta b$. During an iterative optimization process $A_k(\text{m}_\text{k})\Delta x_k = d_k$, $k = 1, 2, \ldots$, iter _ max, we calculate the least-square solution Δx_k for the kth iteration and update the model by

In this formulation, $A(m) = L^*J(m)$ and $d = L^*\Delta b$. The iterative inversion proceeds by solving $A_k(\text{m}_\text{k})\Delta x_k = d_k$, $k = 1, 2, \ldots$, iter _ max for the least-squares solution Δx_k at each step k, followed by an update of the model:

$$m_{k+1} = m_k + \Delta m. \tag{6.10}$$

After obtaining the S-wave velocity model via surface wave inversion, t_0 is estimated by applying the FMM to compute the S-wave traveltime from the sensors to the source, following Eq. 6.1.

6.2.4 AST in Deep Tunnels

Once t_0 is known, the P-wave traveltime is simply the difference between the observed arrival time and t_0. This allows the objective misfit function for traveltime tomography to be established as:

$$E(\boldsymbol{c}) = \frac{1}{2}\int_{\partial\Omega}\left[\boldsymbol{T}(\boldsymbol{c},\boldsymbol{r}) - \boldsymbol{T}^{\text{obs}}(\boldsymbol{r})\right]^2 \mathbf{d}\boldsymbol{r}. \tag{6.11}$$

In this formulation, $T^{obs}(r)$ represents the traveltimes measured at receivers r along the acquisition surface $\partial\Omega$. The corresponding modeled traveltimes $T(c, r)$ are obtained by solving the eikonal equation for the velocity field c.

$$F(\boldsymbol{t}, \boldsymbol{c}) = [\nabla \boldsymbol{t}(\boldsymbol{x})]^2 - \frac{1}{c^2(x)} = 0. \tag{6.12}$$

Here, x denotes an arbitrary point within the domain Ω between the blasting sources and sensors, and the traveltime at source $t(\boldsymbol{x}_{\mathbf{soure}})$ is set to zero. The eikonal equation is solved using the FMM. In the conventional formulation, the objective function can be expanded around c_0 via the Gaussian expansion method.

$$E(\boldsymbol{c}) \approx E(\boldsymbol{c}_0) + \left[\frac{\partial \boldsymbol{E}(\boldsymbol{c})}{\partial \boldsymbol{c}}_{c=c_0}\right]^{\mathbf{T}} \delta c + \frac{1}{2}\delta \boldsymbol{c}^{\mathbf{T}} \frac{\partial^2 \boldsymbol{E}(\boldsymbol{c})}{\partial \boldsymbol{c} \partial \boldsymbol{c}^{\mathbf{T}}}\bigg|_{c=c_0} \delta c. \tag{6.13}$$

The gradient refers to the first derivative of the objective function with respect to c, and the Hessian matrix is its second derivative, as shown in Eq. 6.13.

$$g = \frac{\partial \boldsymbol{E}(\boldsymbol{c})}{\partial \boldsymbol{c}} \tag{6.14}$$

$$H = \frac{\partial^2 \boldsymbol{E}(\boldsymbol{c})}{\partial \boldsymbol{c} \partial \boldsymbol{c}^{\mathbf{T}}} \tag{6.15}$$

Setting the gradient to zero ($g = 0$) provides the necessary condition for minimizing the objective function, leading to:

$$H|_{c0}\, \delta c = -g|_{c0} \tag{6.16}$$

$$\delta c = -(\boldsymbol{H}_{c0})^{-1} g\,|_{c0} \tag{6.17}$$

We can update the velocity model

$$\boldsymbol{c}_{n+1} = \boldsymbol{c}_n - (\boldsymbol{H}|_{\boldsymbol{cn}})^{-1} g\,|_{\boldsymbol{cn}}. \tag{6.18}$$

The model is updated to c_{n+1} and used as the initial guess for the following iteration. Given the difficulty of inverting the Hessian matrix $\boldsymbol{H}$ and the Fréchet matrix g, the adjoint-state method is adopted to compute the gradient without requiring Fréchet derivatives. This is achieved by extending the original objective misfit function (Eq. 6.10) through the inclusion of a Lagrangian term:

$$\begin{aligned} L(\boldsymbol{c}, \boldsymbol{t}, \boldsymbol{\lambda}) = &\ \frac{1}{2}\int_{\partial\boldsymbol{\Omega}} \left[\boldsymbol{t}(\boldsymbol{r}) - \boldsymbol{T}^{\mathbf{obs}}(\boldsymbol{r})\right]^2 \mathbf{d}\boldsymbol{r} \\ & - \frac{1}{2}\int_{\boldsymbol{\Omega}} \boldsymbol{\lambda}(\boldsymbol{x})[\nabla \boldsymbol{t}(\boldsymbol{x})]^2 - \frac{1}{c^2(x)} \mathbf{d}\boldsymbol{\Omega} \end{aligned} \tag{6.19}$$

where $\boldsymbol{t}(\boldsymbol{r})$ is the traveltime field satisfying the eikonal equation at receiver $\boldsymbol{r}$ and $\lambda(\boldsymbol{x})$ is the adjoint-state parameter (Lagrangian multiplier) that imposes the constraint $\boldsymbol{t}(\boldsymbol{r}) = \boldsymbol{T}^{\mathbf{obs}}$ at the optimal point. The variables $\boldsymbol{t}(\boldsymbol{r})$, $\lambda(r)$, and $\boldsymbol{c}(\boldsymbol{x})$ vary independently during the search for the minimum. Upon convergence, they obey the relations in Eqs. 6.20–6.22.

$$\frac{\partial L}{\partial \lambda} = 0 \tag{6.20}$$

$$\frac{\partial L}{\partial t} = 0 \tag{6.21}$$

$$\frac{\partial L}{\partial c} = -\int_{\Omega} \frac{\lambda}{c^3(x)} \mathbf{dx} = \frac{\partial E}{\partial c}. \tag{6.22}$$

We can obtain the following equation by Eq. 6.21

$$\frac{\partial L}{\partial t} = \int_{\partial \Omega} \left[t(\boldsymbol{r}) - \boldsymbol{T}^{\mathbf{obs}}(\boldsymbol{r})\right] \mathbf{d}\boldsymbol{r} - \int_{\Omega} \lambda(\boldsymbol{x}) \nabla t(\boldsymbol{x}) \frac{\partial \nabla t}{\partial t} \mathbf{d}\boldsymbol{x} = 0. \tag{6.23}$$

Application of the Gauss formula to the second term in Eq. 6.23 leads to:

$$\frac{\partial L}{\partial t} = \int_{\partial \Omega} \left[t(\boldsymbol{r}) - \boldsymbol{T}^{\mathbf{obs}}(\boldsymbol{r}) - \lambda \boldsymbol{n} \nabla t\right] \mathbf{d}\boldsymbol{r} + \int_{\Omega} \nabla \lambda(\boldsymbol{x}) \cdot \nabla \boldsymbol{T}(\boldsymbol{x}) \mathbf{d}\boldsymbol{x} = 0. \tag{6.24}$$

The condition that the two integral terms in Eq. 6.23 vanish leads to the following relation for λ:

$$\nabla \cdot (-\lambda \nabla \boldsymbol{T}) = 0 \tag{6.25}$$

with boundary conditions

$$(\boldsymbol{n} \cdot \nabla \boldsymbol{T})\lambda = T(\boldsymbol{r}) - \boldsymbol{T}^{\mathrm{obs}}(\boldsymbol{r}). \tag{6.26}$$

Eq. 6.25 is solved for λ over the full model domain using the fast sweeping method (Taillandier et al. 2009), with the boundary condition given by Eq. 6.26. The gradient $\frac{\partial E}{\partial c}$ is subsequently derived from Eq. 6.22, leading to the following model update:

$$\boldsymbol{c}_{n+1} = \boldsymbol{c}_n + \alpha \frac{\partial E}{\partial c}. \tag{6.27}$$

The step length is denoted by α. Iterations are performed until the objective function decreases below a predefined stopping criterion.

6.2.5 Locating MS Sources with Heterogeneous Velocity Models

Source location in deep tunnels employs an objective function based on the L2 norm of double-difference traveltimes (Eq. 6.2). Forward traveltime calculations rely on the FMM and the inverted heterogeneous P-wave velocity model. Minimization is performed using the simplex method. By applying reflection, expansion, contraction, and shrinkage operations (Li et al. 2014), the method systematically replaces higher-value points with lower-value ones until the minimum is attained.

It should be noted that, owing to the linear monitoring geometry inherent in tunnel construction, the inverted heterogeneous velocity models are 2D. Moreover, the source initiation time is not eliminated in the double-difference objective function. Consequently, the simplex geometry in 2D space reduces to a triangle. The workflow of the double-difference source location method based on the simplex approach can be summarized as follows:

(a) The double-difference source location objective function is established as given in Eq. 6.2. Based on the MS monitoring area, three initial points P_1, P_2, and P_3 are selected to form the initial simplex triangle.

(b) Using the current inverted heterogeneous velocity model, the FMM is employed to compute the three objective function values y_1, y_2, and y_3 corresponding to points P_1, P_2, and P_3. The minimum, middle, and maximum values among y_1, y_2, and y_3 are denoted as $y_{\min}$, y_{mid}, and $y_{\max}$, respectively, with their corresponding points identified as $P_{\min}$, P_{mid}, and $P_{\max}$. The reflection, expansion, and contraction points, denoted $P*$, $P**$, and $P***$, and their associated values $y*$, $y**$, and $y***$ are subsequently defined.

(c) The iterative search procedure is initiated. The reflection point $P*$ is evaluated, and its objective function value $y*$ is compared with $y_{\min}$ and y_{mid}. Based on this comparison, a new simplex is generated according to the four transformation operations, reflection, expansion, contraction, and shrinkage, as illustrated in Fig. 6.3. Detailed formulations of these transformations are provided in Li et al. (2014).

(d) Steps (b) and (c) are repeated until the misfit falls below a predefined threshold, at which point the optimal source location is considered found.

(e) When additional blasting sources are incorporated into the traveltime tomography, yielding an updated heterogeneous P-wave velocity model, steps (a) through (d) are repeated to refine the source location.

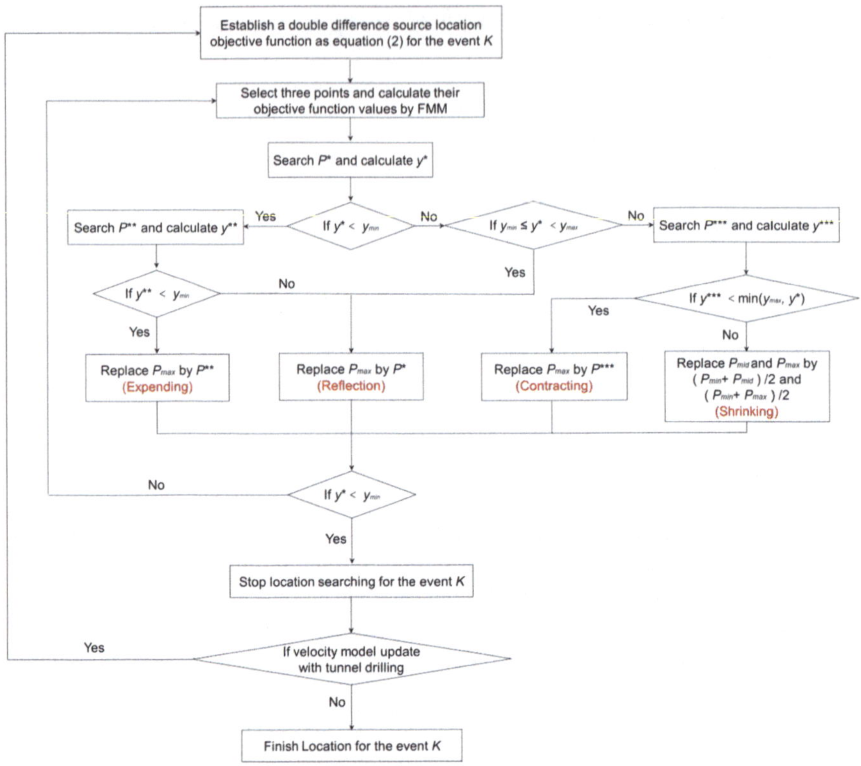

Fig. 6.3 Workflow of double-difference simplex-based MS source location

6.3 Numerical Study

A numerical simulation based on a realistic deep tunnel model is conducted to demonstrate the proposed inversion method for heterogeneous velocity models. Wavefield snapshots are presented to visualize the propagation characteristics of P-waves, S-waves, and surface waves within the tunnel environment. The velocity model, incorporating the effect of the EDZ, is inverted using blasting data. The performance of the proposed MS source location method is then evaluated by locating and analyzing multiple MS events in the vicinity of the tunnel face.

6.3.1 Numerical Modeling of the Blasting Source

Figure 6.4 presents the numerical model, which includes an existing tunnel and two new tunnels under construction oriented perpendicular to the existing tunnel. A convolutional perfectly matched layer boundary is applied to the edges of the

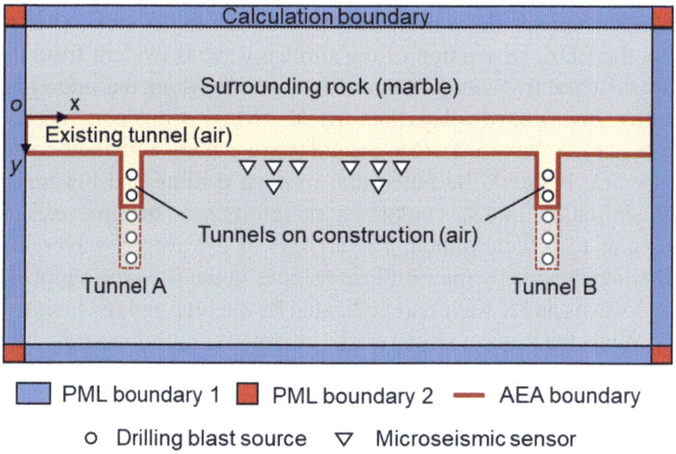

Fig. 6.4 Scheme of the MS source location numerical simulation model6

computational domain to effectively absorb reflections and enable wave propagation simulation within a finite region. High-order finite-difference time-domain methods are employed to solve the viscoelastic wave equation. Attenuation and dispersion effects of viscoelastic waves are accounted for using the generalized standard linear solid. The acoustic-elastic boundary method (Xu et al. 2007) is adopted to handle the free-surface (air) boundary condition.

The two new tunnels are excavated alternately, as depicted in Fig. 6.4. The computational domain spans 100 m in width and 50 m in depth, discretized with a spatial interval of 0.2 m in both the x- and y-directions. The starting points (x, z) of tunnels A and B, defined as the midpoint of each drilling face, are located at (15, 0) and (85, 0), respectively. The MS monitoring network is installed centrally between the two tunnels and comprises eight sensors positioned at the following coordinates: (35, 1), (40, 1), (40, 4), (45, 1), (55, 1), (60, 1), (60, 4), and (65, 1). Drilling and blasting sources are placed along the central axis of each tunnel at 2 m intervals. Each blasting source is modeled as a Ricker wavelet with a dominant frequency of 120 Hz. The surrounding rock is characterized by background S-wave and P-wave velocities of 3500 m/s and 5250 m/s, respectively, and a density of 2.5 g/cm^3. Within a 10 m zone around the tunnel, the S-wave velocity decreases from 3500 to 2500 m/s, while the P-wave to S-wave velocity ratio remains constant at 1.5. The total simulation duration is 50 ms, with a sampling interval of 0.02 ms.

Figure 6.5 presents snapshots of the wavefield generated by the first blasting source, initiated at the central axis of tunnel A. Due to the difference in propagation velocities, the P- and S-waves gradually separate over time. A surface wave (Rayleigh wave) traveling along the tunnel sidewall becomes observable after t_0+13.5 ms. Its propagation velocity is slightly lower than that of the S-wave, typically approximately 0.9 times the S-wave velocity. At t_0+21 ms, a P-to-S converted wave appears, resulting from the reflection of the P-wave at the tunnel sidewall. The S-wave propagates more rapidly in deeper regions along the y-direction, as the intact rock exhibits a higher velocity than the EDZ. Dispersion of the surface wave is evident from t_0+13.5 ms onward, with different frequency components traveling along the sidewall at distinct phase velocities. Figure 6.6 displays the vertical velocity component recorded by four sensors located at (35, 1), (45, 1), (55, 1), and (65, 1), along with their corresponding frequency spectra. It should be noted that in most drilling and blasting tunneling scenarios, the initiation time t_0 is unknown; therefore, t_0 on the time axes in Figs. 6.5 and 6.6a remains to be determined. The arrivals of P-waves, S-waves, and surface waves are distinguishable by their differing slopes in the waveform plot. The arrival times of the P-waves and S-waves are indicated by the blue and red lines in Fig. 6.6a, respectively. While the P-wave arrival can be obtained straightforwardly for all traces, the S-wave arrival is clearly identifiable only for sensor No. 4; for sensors Nos. 1, 2, and 3, accurate S-wave picking is hindered by overlapping phases. Surface wave dispersion becomes evident in the extended waveforms recorded at greater offsets. The dominant frequency of the recorded signals ranges from 150 to 170 Hz (Fig. 6.6b), which is consistent with the frequency characteristics of actual blasting sources in tunnel environments.

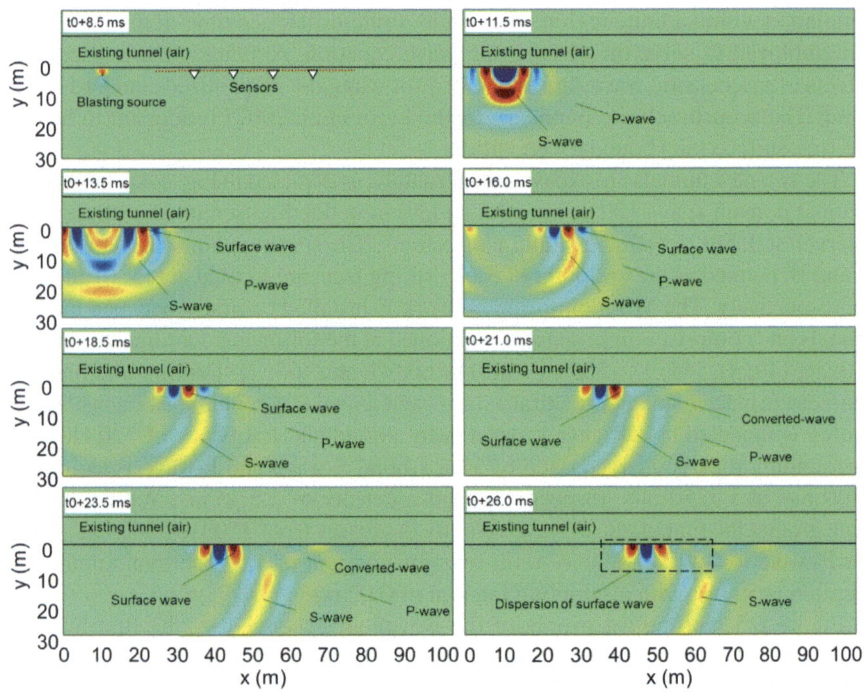

Fig. 6.5 Snapshots of the first blasting source at tunnel A

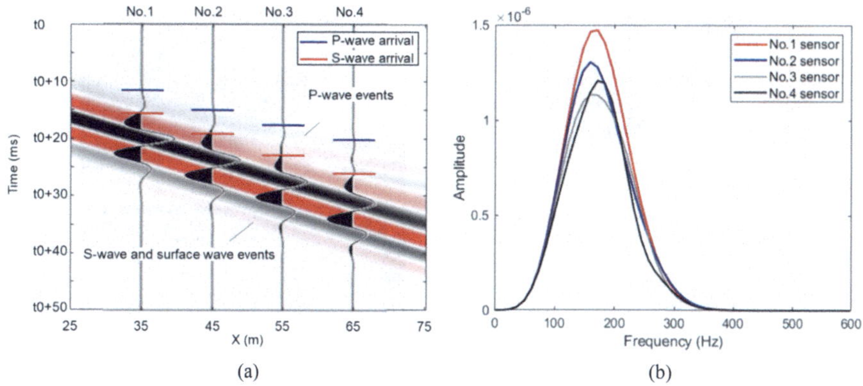

Fig. 6.6 Simulated recordings of four sensors at the (35, 1), (45, 1), (55, 1), and (65, 1) **a** wave profiles and **b** corresponding frequency spectra

6.3.2 Determining t_0 *of the Blasting Source*

The phase-shift method (Park et al. 1998) is applied to compute the dispersion curves from the four recorded traces. The resulting dispersion curve is shown in Fig. 6.7a. Due to the limited number of sensors, spatial aliasing is observed in the calculated dispersion image. For comparison, a synthetic dispersion curve is computed assuming 11 evenly spaced sensors located from (25, 1) to (75, 1), as presented in Fig. 6.7b. The fundamental mode dispersion curves obtained from both configurations exhibit strong agreement, and the energy distribution in the dispersion images aligns well with the theoretical solutions (indicated by white stars in Fig. 6.7). This comparison confirms the accuracy of the surface wave modeling and demonstrates that the fundamental mode dispersion curve can be successfully extracted using as few as four linearly arranged monitoring sensors.

The dispersion curve extracted from Fig. 6.7a is inverted using the proposed improved perturbational inversion method, with the inversion process illustrated in Fig. 6.8. The initial S-wave velocity model is assumed to vary linearly from 2500 to 2700 m/s (blue dashed line in Fig. 6.8c), and its corresponding dispersion curve is shown in Fig. 6.8a. A noticeable discrepancy exists between the observed and initial dispersion curves. The inversion seeks to reduce this misfit by iteratively updating the S-wave velocity model. Throughout the inversion, the P-wave to S-wave velocity ratio is fixed at 1.5, and the density is held constant. The misfit decreases rapidly and converges to a normalized value of 0.14 after 120 iterations, after which no further improvement is observed. The inverted S-wave profile reveals a distinct low-velocity zone extending from the surface to a depth of 10 m, consistent with the true S-wave model. Moreover, the dispersion curve computed from the final inverted model closely matches the observed data.

The resulting layered S-wave velocity model is presented in Fig. 6.9a. Contours of equal traveltime are calculated using the FMM. Ray paths from the sensor to

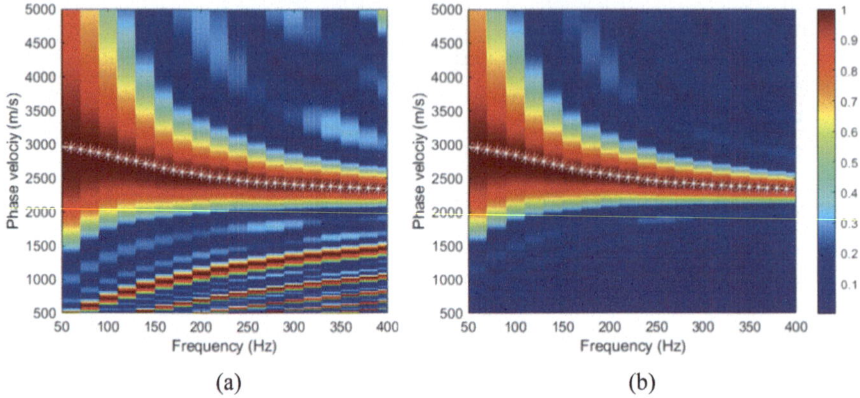

Fig. 6.7 Comparison of dispersion curves from blasting source data: **a** curve calculated from four field recordings; **b** curve derived from simulated sensor array along the red dashed line in Fig. 6.5. Theoretical dispersion solutions are marked by white stars

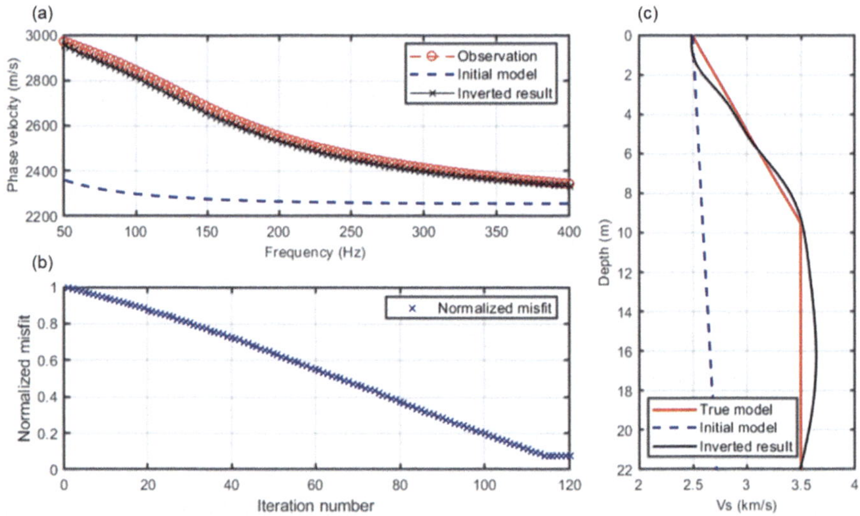

Fig. 6.8 Dispersion curve inversion using the improved perturbational method: **a** Comparison of observed and calculated dispersion curves at the 120th iteration; **b** iterative decrease in the misfit function; **c** inverted versus true S-wave velocity model at iteration 120

the source are obtained by tracing the normal direction to each isochrone. These ray-tracing results indicate that wave propagation paths differ substantially between the inverted S-wave model and a homogeneous model. In the inverted model, the low-velocity EDZ causes the ray paths to curve and lengthen, whereas in the homogeneous model, the paths remain straight. The travel times from the four sensors to the source, computed using the actual, inverted, and homogeneous models, are

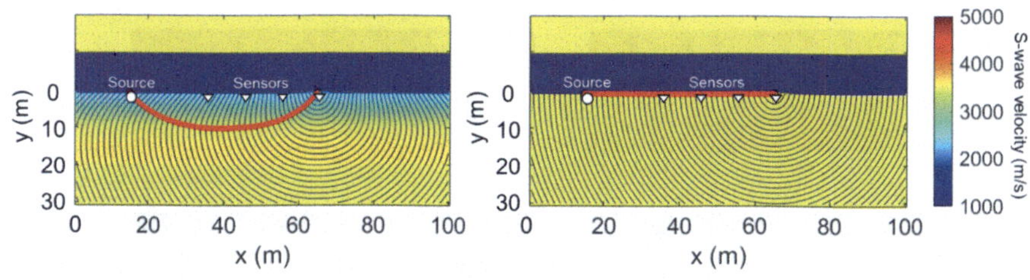

Fig. 6.9 Backpropagation of S-waves from sensor 4 to the first blasting source in tunnel A, illustrating ray paths calculated from **a** the inverted S-wave velocity model and **b** a homogeneous S-wave velocity model

Table. 6.1 Calculation of the drilling blast source of the numerical study

Number of sensors	S-wave arrival time (ms)	Calculated traveling time of real S-wave velocity model (ms)	Calculated traveling time of inverted S-wave velocity model (ms)	Calculated traveling time by homogenous S-wave velocity model (ms)	Calculated time of actual S-wave velocity model (ms)	Calculated time of inverted S-wave velocity model (ms)	Calculated time of homogenous S-wave velocity model (ms)
1	15.86	7.80	8.01	5.71	8.06	7.85	10.15
2	19.31	11.27	11.59	8.57	8.04	7.72	10.74
3	22.42	14.40	15.63	11.43	8.02	7.79	10.99
4	25.33	17.29	17.50	14.28	80.4	7.83	11.05
Average	–	–	–	–	8.04	7.79	10.70

summarized in Table 6.1. The traveltimes obtained from the inverted model are in close agreement with those from the actual model. In contrast, traveltimes derived from the homogeneous model are significantly shorter due to the neglect of the EDZ. The blasting initiation time t_0 is estimated using the S-wave arrival times and Eq. 6.2. As reported in Table 6.1, the error in t_0 obtained from the inverted S-wave model is approximately 0.2 ms, substantially smaller than the 2–3 ms error associated with the homogeneous model. These results indicate that t_0 estimated using the inverted S-wave model can be reliably used as the trigger time of the blasting source.

6.3.3 *P-Wave Traveling Time Tomography While Drilling*

Following the determination of the blasting source trigger time t_0, P-wave traveltime tomography is carried out. The absolute P-wave traveltime is obtained by combining t_0 with the manually or automatically picked P-wave arrival times (e.g., the blue line

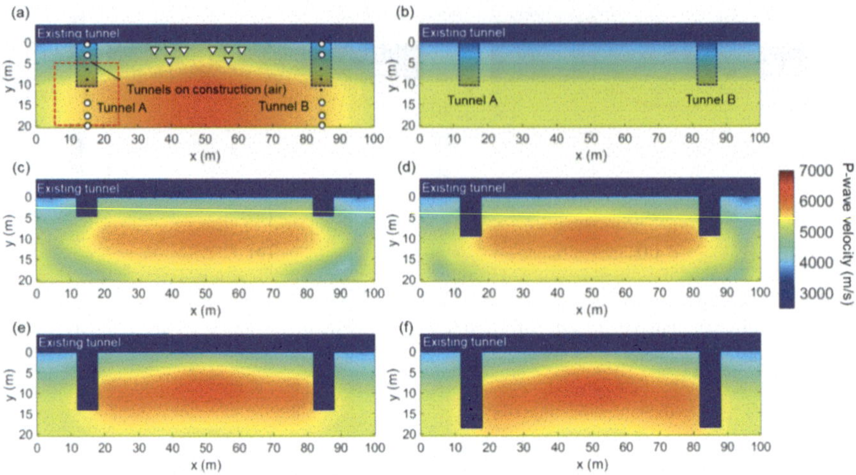

Fig. 6.10 Inverted P-wave velocity models at different excavation stages: **a** reference true model; **b** initial homogeneous model; **c–f** tomography results at drilling depths of 5, 10, 14, and 18 m

in Fig. 6.6a). In the simulation, blasting sources are positioned along the central axes of the two tunnels at 2 m intervals. Traveltime data from the eight sensors deployed between the tunnels are used to invert the P-wave velocity structure. The results of the P-wave traveltime tomography at successive stages of excavation are presented in Fig. 6.10.

Figure 6.10a shows the true P-wave velocity model, which includes low-velocity zones corresponding to the EDZ around the existing tunnel and the two new tunnels. Regions farther from the tunnel sidewalls exhibit higher P-wave velocities. The initial velocity model for tomography is derived from the previously inverted S-wave velocity profile. Tomographic inversions using the adjoint-state method are performed at four excavation depths: 5 m, 10 m, 14 m, and 18 m.

As illustrated in Fig. 6.10c–f, the inverted P-wave velocity model is progressively refined toward the true model. In the first stage (2 m depth, Fig. 6.10c), velocity updates are primarily concentrated in the vicinity of the existing tunnel, where ray paths from the initial blasting sources are concentrated. Owing to the symmetry of the sensor layout and the true velocity model, the inverted models remain approximately symmetric throughout the process. By the second stage (Fig. 6.10d), the spatial distribution of the velocity improves, although noticeable discrepancies from the true model persist. After the third and fourth stages (Figs. 6.10e, f), the inverted models closely approximate the true velocity distribution.

A detailed comparison of the velocity profiles at selected x-positions (20, 30, 40, and 50 m) is provided in Fig. 6.11. These profiles demonstrate that the adjoint-state traveltime tomography effectively recovers a reasonable heterogeneous velocity model as excavation progresses. With the inclusion of additional blasting sources, the inverted P-wave velocities increasingly converge toward the true values. Some

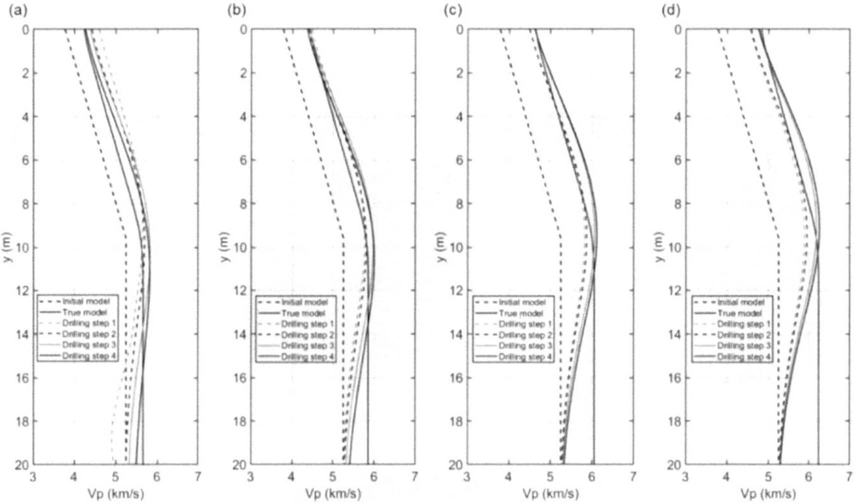

Fig. 6.11 Inverted p-wave velocity comparison: **a** $x = 20$ m, **b** $x = 30$ m, **c** $x = 40$ m, and **d** $x = 50$ m

residual error remains in the central region of the monitoring area (approximately $x = 50$, $y > 12$ m) due to limited ray path coverage. However, this discrepancy does not compromise the accuracy of MS source localization, as waves generated by MS events near the tunnel also do not traverse this region en route to the sensors.

6.3.4 MS Source Location

The location accuracy of the proposed method is evaluated using 64 MS sources uniformly distributed within the red square region in Fig. 6.10a, spanning x-coordinates from 5 to 19 m and y-coordinates from 8 to 22 m. Forward-simulated arrival times are computed by solving the eikonal equation using the FMM. Location errors obtained using various homogeneous velocity models and inverted velocity models from different excavation stages are summarized in Fig. 6.12.

Figure 6.12a presents the location errors when the true velocity model is used, effectively representing the baseline performance of the double-difference simplex method in the absence of velocity model error. The errors are consistently small, confirming the intrinsic accuracy of the location algorithm. Consequently, when inverted or homogeneous velocity models are employed, the observed errors primarily reflect inaccuracies in the velocity model itself.

The results obtained using homogeneous velocity models (4500, 5000, and 5500 m/s) exhibit substantial location errors, as shown in Fig. 6.12b–d. These errors are largely governed by the discrepancy between the assumed homogeneous velocity and the true heterogeneous field, particularly within the EDZ. For instance, at a

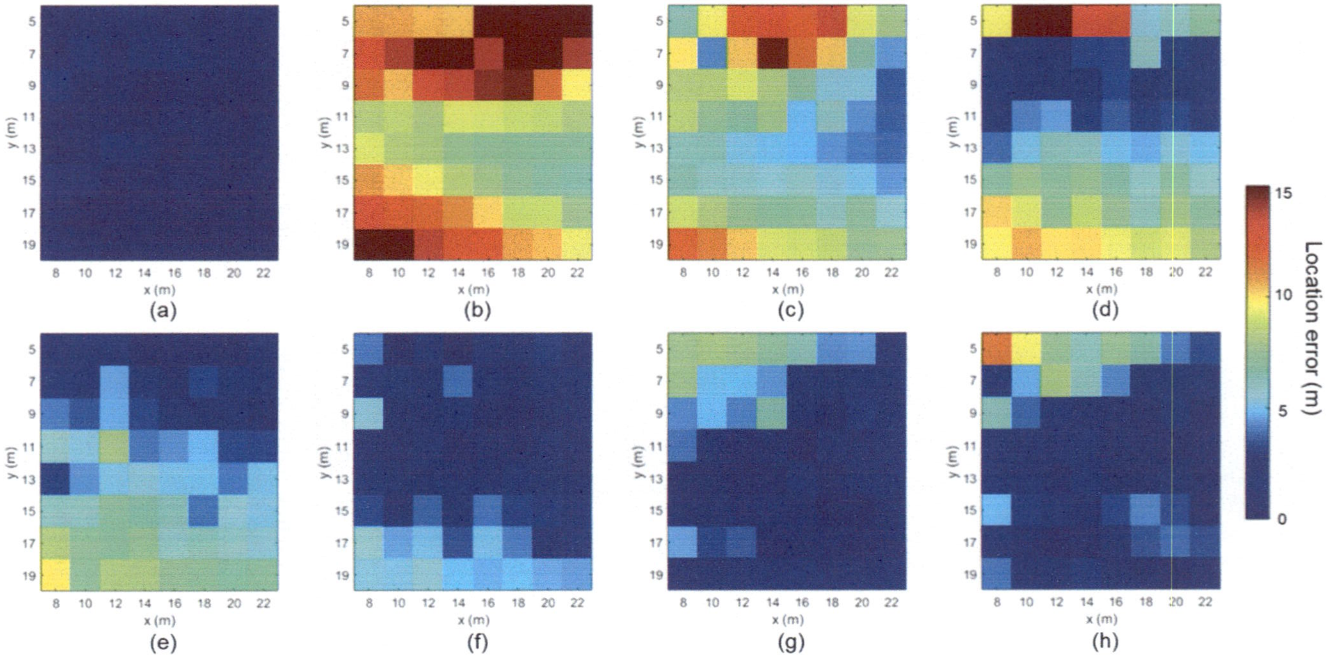

Fig. 6.12 MS source location errors (red dashed square in Fig. 6.10a) derived from different P-wave velocity inputs: **a** true model; **b** homogeneous 4500 m/s; **c** homogeneous 5000 m/s; **d** homogeneous 5500 m/s; **e** Step 1 inverted model (Fig. 6.10c); **f** Step 2 inverted model (Fig. 6.10d); **g** Step 3 inverted model (Fig. 6.10e); **h** Step 4 inverted model (Fig. 6.10f)

depth of 9 m, the model with 5500 m/s yields relatively smaller errors, as this value approximates the average velocity at that depth. However, at other depths, the inherent heterogeneity of the true model leads to pronounced increases in error.

Notably, the location accuracy improves progressively when the updated inverted velocity models are used (Fig. 6.12e–h). The inverted models correspond to blasting sources at drilling depths of 5 m, 10 m, 14 m, and 18 m, respectively. As a result, the locations of MS events near these depths are estimated with high precision, with most errors falling below 2 m. In contrast, areas with lower ray path coverage in the traveltime tomography exhibit larger residuals. Overall, the location accuracy achieved with the inverted velocity models is substantially higher than that obtained with any homogeneous model, and it continues to improve as more blasting sources are incorporated into the P-wave velocity tomography.

6.4 Case Study

MS monitoring was conducted in two newly excavated tunnels at a depth of 2400 m within the Jinping Underground Laboratory Phase II (CJPL-II) to provide early warning of rockburst hazards. Based on the analysis of drilling and blasting source data, a heterogeneous P-wave velocity model of the monitored area was inverted. Blasting events with known locations were subsequently used to validate the performance of the proposed location method.

6.4.1 Background

CJPL-II is one of the deepest underground laboratories on the Yalong River in Sichuan Province, China (Zheng et al. 2021). The CJPL-II underground laboratory has a maximum depth of 2525 m and is situated in marble with high strength and brittleness. Field measurements report a maximum principal stress of 69.20 MPa in the surrounding rock mass (Jiang et al. 2021). Based on previous studies, rockburst events frequently occur during the excavation of tunnels and underground laboratories (Feng et al. 2015). Fig. 6.13a illustrates the layout of the eight caverns of the CJPL-II project and their connections to the access tunnels. Within the area indicated by the dashed square, two new tunnels were excavated simultaneously in the No. 2 auxiliary tunnel using the drilling and blasting method. To enable efficient monitoring, the sensor network was installed centrally between the two tunnels, allowing MS events near the working faces of both tunnels to be recorded by a single monitoring system. A SinoSeism MS monitoring system equipped with a 32-bit A/D acquisition unit (Chen et al. 2021) was deployed for this purpose. The monitoring layout comprised eight sensors, as shown in Fig. 6.13c. The nearest sensor was positioned 20 m from the tunnel face to prevent damage from blasting waves. Owing to the planar geometry

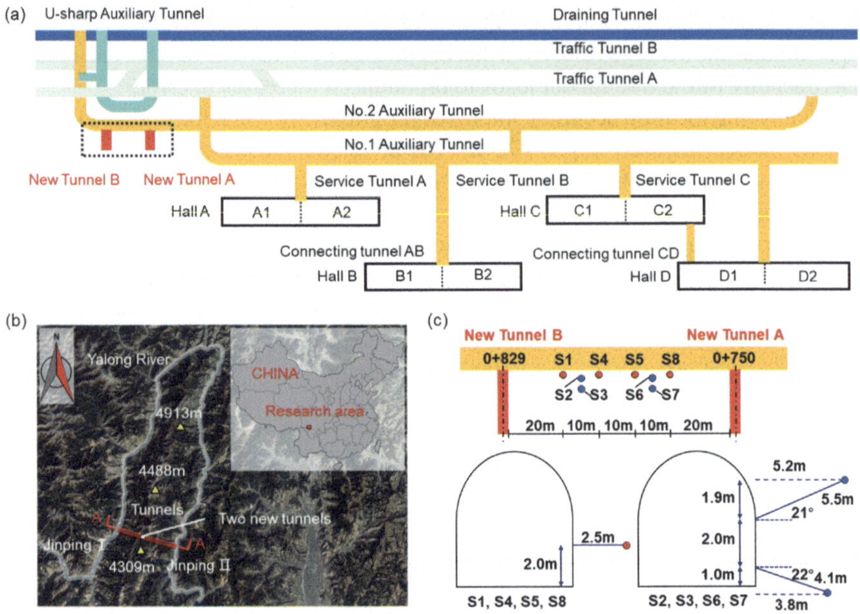

Fig. 6.13 Background of CJPL-II and the MS monitoring setup of two new tunnels: **a** the layout of CJPL-II, **b** the location of Jinping China, and **c** the MS monitoring setup

of the monitoring network and the target region, both velocity model inversion and MS source location were simplified to a two-dimensional problem.

6.4.2 Inversion of the Velocity Model

Blasting data were acquired to invert the velocity models of the monitored area. The coordinates of the 14 blasting sources, seven in tunnel A and seven in tunnel B, are listed in Tables 6.2 and 6.3, respectively. The geological conditions encountered at the drilling faces during excavation are illustrated in Fig. 6.14. Most of the tunnel faces consisted of grayish-white fine-crystalline marble, exhibiting a blocky structure and generally good integrity. In certain areas, the rock mass appeared rust-yellow due to ferromanganese infilling. The fractures were dry, fresh, and unweathered, and no audible sound was noted during photography. According to the MS monitoring results, a minor rockburst risk was identified at tunnel B 0+4.5, while the risk level remained low at all other locations. In response, the section at tunnel B 0+4.5 was treated with shotcreting, anchoring, and grouting during construction, and no rockburst occurred throughout the excavation process. For surface wave dispersion analysis, four sensors positioned at the same elevation (S_1, S_4, S_5, and S_8 in

Table. 6.2 Calculation of t_0 of the different drilling blast sources in the field case (Tunnel A)

Number of tunnel	Coordinate of known blasting source (m)		S-wave arrival time of no.1 sensor (year-month-day h:min:s)	S-wave absolute traveling time of no. 1 sensor (ms)	Time of drilling blast source in tunnel A (year-month-day h:min:s)
Tunnel A	A1	(750, 2)	2021-9-7 18:56:35.60755	16.08	2021-9-7 18:56:35.59148
	A2	(750, 4.5)	2021-9-15 01:09:13.27175	15.82	2021-9-15 01:09:13.25593
	A3	(750, 7)	2021-9-20 00:04:14.56410	15.59	2021-9-20 00:04:14.54851
	A4	(750, 10.5)	2021-9-23 20:52:23.54895	15.50	2021-9-23 20:52:23.5334
	A5	(750, 14)	2021-9-30 13:23:56.23615	15.75	2021-9-30 13:23:56.22040
	A6	(750, 17.5)	2021-10-3 21:34:45.70745	16.16	2021-10-3 18:56:35.6912
	A7	(750, 20)	2021-10-5 14:03:42.57210	16.44	2021-10-5 18:56:35.55566

Fig. 6.13c) were used, while all eight sensors were incorporated into the P-wave velocity tomography.

6.4.2.1 Data Processing of the Blasting Signal

The signal processing results for the first blasting source in tunnels A and B are presented in Fig. 6.15. The raw waveforms shown in Fig. 6.15a, e exhibit significant low-frequency noise in traces S_1, S_4, and S_8, attributed to electrical interference from the sensors. To suppress this noise, a bandpass filter is applied. The filtered waveforms in Fig. 6.15b, f reveal multiple distinct pulses, indicating that several independent blasts occurred and can be analyzed separately. To further examine the waveform characteristics, the time window corresponding to the first pulse is extracted, as shown in Fig. 6.15c, g. The first arrival of the P-wave is clearly identifiable across all traces. As the source–receiver offset increases, the P- and S-waves gradually separate. Consequently, accurate identification of the first S-wave arrival is challenging in near-offset traces. However, at greater offsets, such as from S_1 to S_4 in the tunnel A dataset and from S_5 to S_8 in the tunnel B dataset, the S-wave onset becomes clearly distinguishable. Notably, sensor pairs at the same horizontal positions (S_2 and S_3, S_6 and S_7) exhibit highly similar waveforms and arrival times, confirming the validity of the two-dimensional assumption used in the inversion and location procedures.

Table. 6.3 Calculation of t_0 of the different drilling blast sources in the field case (Tunnel B)

Number of tunnel	Coordinate of known blasting source (m)		S-wave arrival time of no.8 sensor (year-month-day h:min:s)	S-wave absolute traveling time of no.8 sensor (ms)	Time of drilling blast source in tunnel B (year/month/day h:min:s)
Tunnel B	B1	(829, 2)	2021-9-1 23:19:51.64450	15.92	2021-9-1 23:19:51.62853
	B2	(829, 4.5)	2021-9-6 18:41:53.89455	15.68	2021-9-6 18:41:53.87887
	B3	(829, 9)	2021-9-12 22:55.12.36590	15.35	2021-9-12 22:55.12.35055
	B4	(829, 12.5)	2021-9-16 00:20:43.35950	15.47	2021-9-16 00:20:43.34403
	B5	(829, 15)	2021-9-18 23:50:23.44895	15.72	2021-9-18 23:50:23.43323
	B6	(829, 18)	2021-9-21 18:10:32.12355	16.08	2021-9-21 18:10:32.10747
	B7	(829, 20)	2021-9-28 19:30:53.35610	16.30	2021-9-28 19:30:53.33980

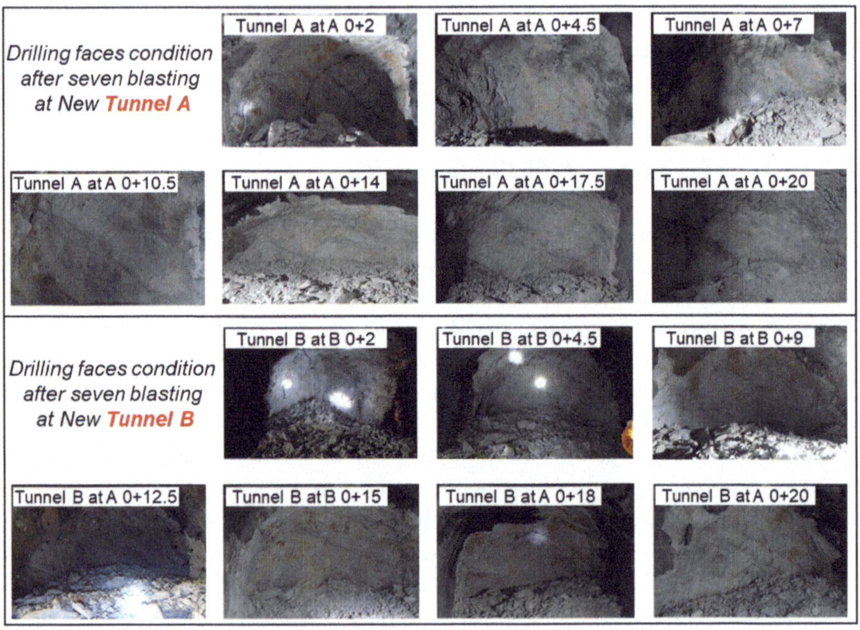

Fig. 6.14 Drilling face conditions after blasting of two new tunnels

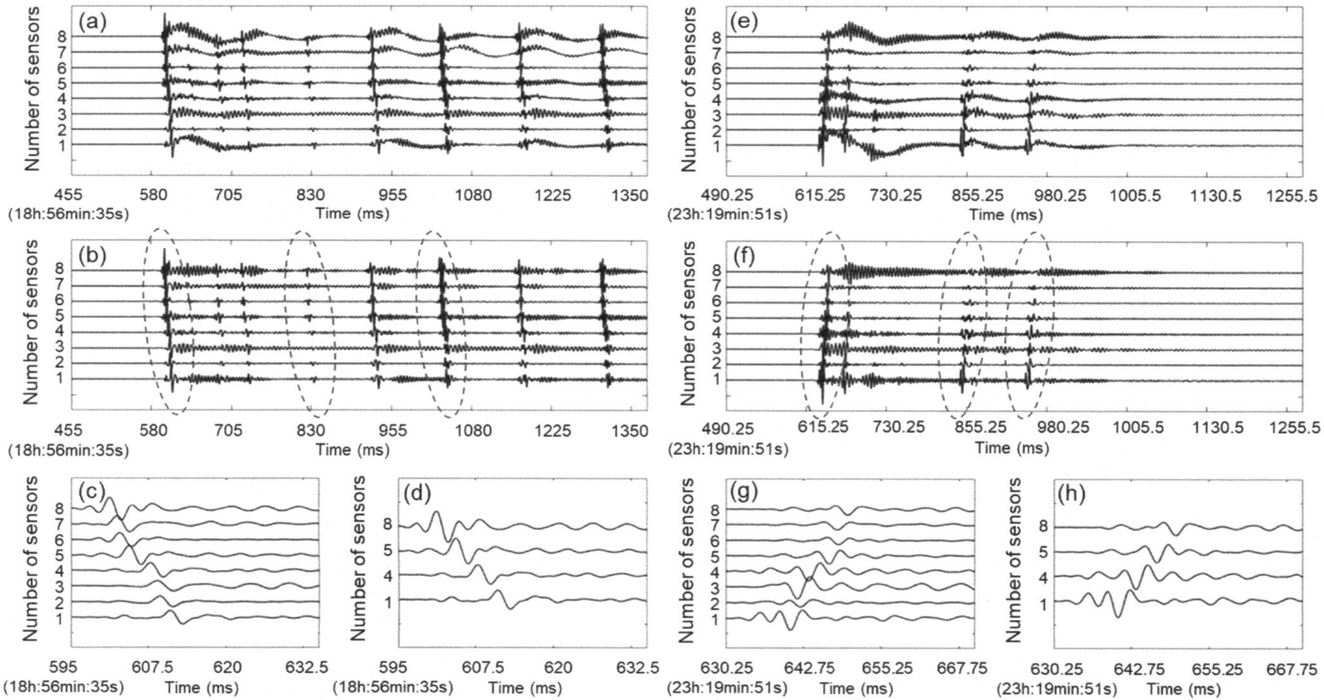

Fig. 6.15 Processing of blasting data from the first source in tunnels A and B. **a** Raw data of blasting source A1; **b** filtered signal of A1; **c** time window containing the first pulse extracted from (**a**); **d** aligned waveforms of channels S_1, S_4, S_5, and S_8 for the first pulse of A1; **e** raw data of blasting source B1; **f** filtered signal of B1; **g** first-pulse time window extracted from (**e**); **h** corresponding waveforms of channels S_1, S_4, S_5, and S_8 for the first pulse of B1

6.4.2.2 Dispersion Analysis

Dispersion curves were extracted from the blasting source data of tunnels A and B, as shown in Fig. 6.15d, h. The two dispersion energy distributions presented in Fig. 6.16a, b exhibit comparable resolution, reflecting a consistent velocity–depth profile. The dispersion patterns evident in Fig. 6.16 indicate variations in seismic velocity with depth within the surrounding rock. Both curves reveal a lower-velocity zone at higher frequencies and a higher-velocity region near 175 and 50 Hz.

The proposed improved perturbational inversion method was applied using the averaged dispersion curve from the two datasets to derive the S-wave velocity model. The initial model assumed a linear velocity increase from 3200 to 3500 m/s (blue dashed line in Fig. 6.17c). The misfit between the observed and simulated dispersion curves decreased rapidly, converging to a normalized value of 0.11 after 100 iterations (Fig. 6.17a, b). The inverted S-wave profile reveals a distinct low-velocity zone extending from the surface to a depth of approximately 10 m, consistent with the presence of an EDZ. As shown in Fig. 6.17c, the inverted velocity–depth curve suggests an EDZ thickness of approximately 2.8 m, with intact high-velocity rock encountered at depths near 9 m. The low-velocity region is interpreted as indicative of a fractured rock mass or developed discontinuities.

Backward ray tracing from sensor S_1 to the sources was conducted to estimate the initiation time t_0 of the blasting sources in tunnel A. The FMM was employed to compute traveltimes and corresponding isochrones, as illustrated in Fig. 6.18. Propagation paths within the heterogeneous velocity model were obtained by tracking the direction normal to the isochrones. Using the S-wave arrival times and traveltimes recorded by sensors S_1 and S_8, the t_0 values for the blasting sources in tunnels A and B were calculated according to Eq. 6.1. The resulting estimates are summarized in Table 6.2.

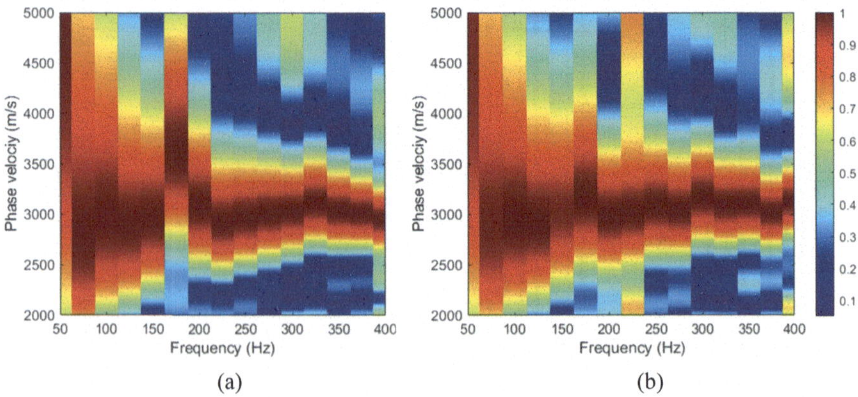

Fig. 6.16 Dispersion curve extracted from sensors S_1, S_4, S_5, and S_8 for **a** tunnel A blasting source and **b** tunnel B blasting source

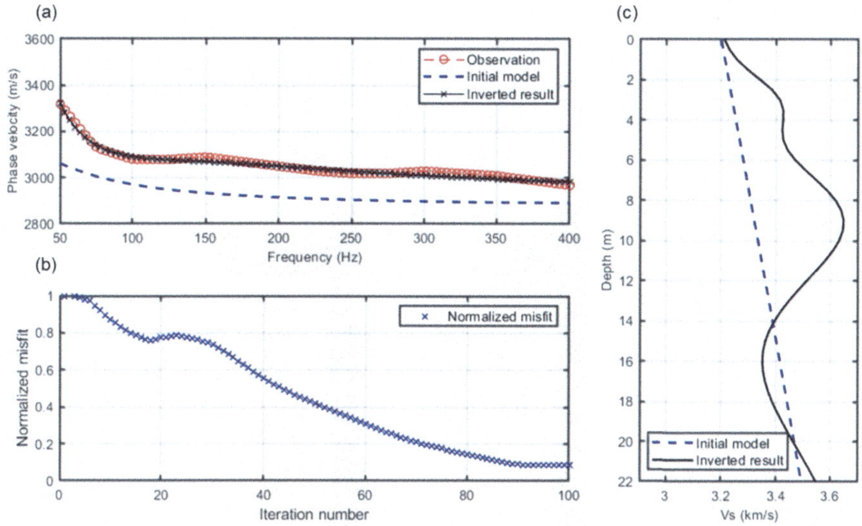

Fig. 6.17 Field data dispersion curve inversion: **a** Comparison of measured and computed dispersion curves at iteration 100; **b** misfit evolution throughout the inversion; **c** Inverted S-wave velocity structure at the 100th iteration

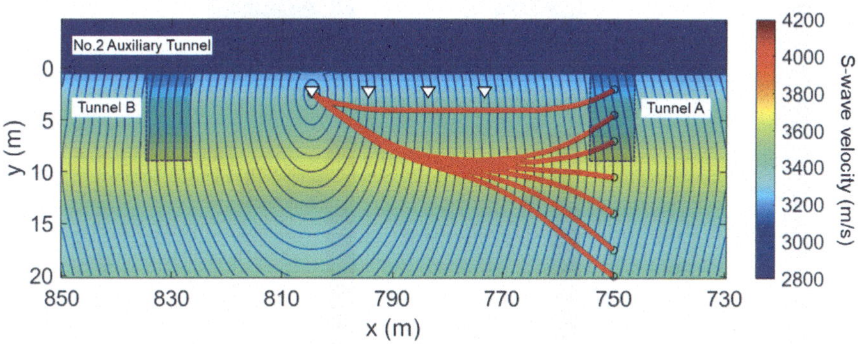

Fig. 6.18 Determining t_0 of blasting source A1 by backpropagation of the s-wave from S_1 sensors

6.4.2.3 P-Wave Traveling Time Tomography

The calculated initiation times t_0 of the blasting sources were subsequently employed to obtain the P-wave travel times. The P-wave travel times for each sensor–source pair are summarized in Table 6.3. The initial model for P-wave travel time tomography, shown in Fig. 6.19a, assumes a linear velocity increase from 5000 to 5500 m/s over a depth range of 0–10 m. The inversion was performed in four stages, incorporating different sets of blasting sources at each step. The resulting inverted P-wave velocity models are presented in Fig. 6.19b–e.

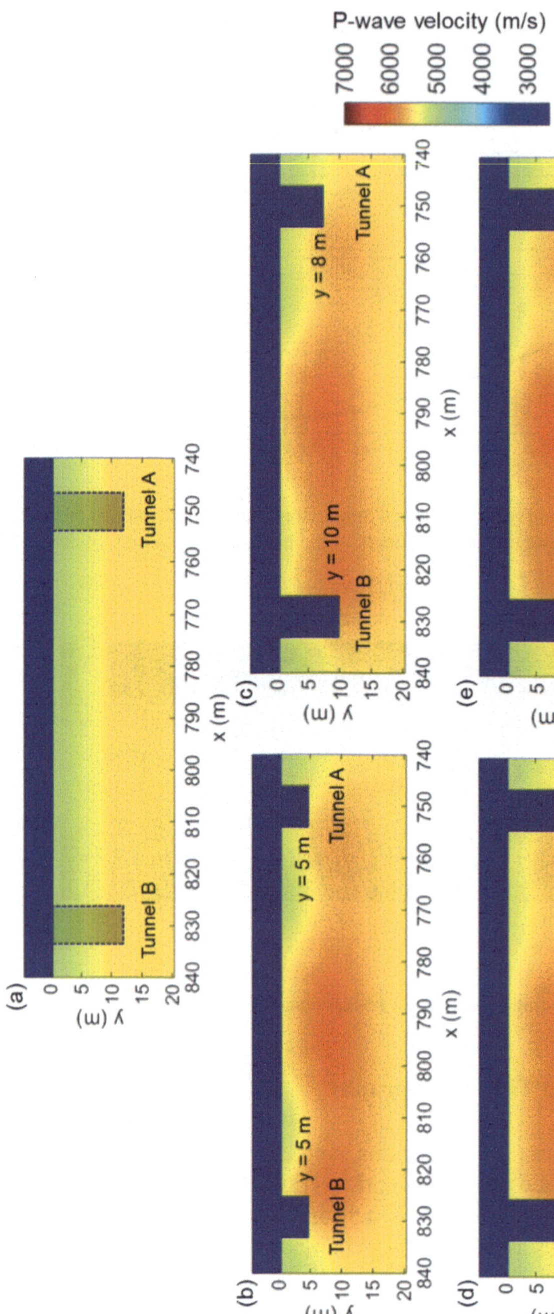

Fig. 6.19 Sequential inversion of P-wave velocity models using increasing numbers of blasting sources: **a** initial model; **b** with sources A1, A2, B1, B2; **c** with sources A1–A3 and B1–B3; **d** with sources A1–A5 and B1–B5; **e** with all sources A1–A7 and B1–B7

All inverted models successfully delineate the EDZ. A region of elevated velocity is observable between depths of 5 m and 10 m, indicating the presence of intact rock. In the models obtained from the third and fourth inversion steps, a lower-velocity zone appears between 14 m and 20 m. Furthermore, the velocity models shown in Fig. 6.19c–e consistently indicate that the surrounding rock near tunnel A exhibits lower velocities than that near tunnel B.

6.4.3 Validation by the Blasting Source Location

Although no significant rockburst occurred during tunnel excavation, a considerable number of MS events were recorded throughout the process. To validate the proposed location method, 14 blasting sources with precisely known locations were used as test data. Location accuracy was evaluated using four inverted velocity models corresponding to different drilling stages (Fig. 6.19c–e), and the resulting errors are summarized in Fig. 6.20. The majority of location errors fall below 5 m, with deeper sources exhibiting smaller errors.

The results obtained using inverted model 1 (Fig. 6.19c) are represented by the blue line in Fig. 6.20. These errors remain consistently below 4 m throughout the entire drilling process. Inverted model 2 (Fig. 6.19d) yields improved location accuracy at a depth of approximately 9 m, where newly added blasting sources (at depths of 9 m in tunnel A and 7 m in tunnel B) were incorporated into the tomography. A similar trend is observed for inverted model 3 (gray line) in the depth range of 15–18 m and for inverted model 4 (black line) between 18 m and 20 m. In each case, when a blasting source is newly introduced into the tomographic inversion, the location accuracy for that source and its surrounding area is significantly enhanced, with errors typically falling below 2 m.

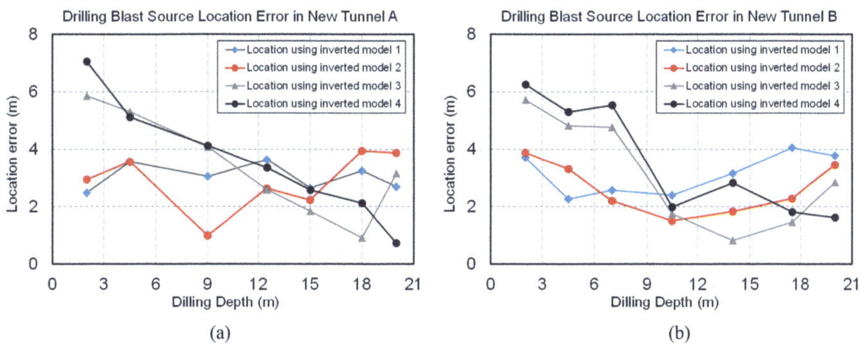

Fig. 6.20 Accuracy of blasting source locations obtained with various inverted velocity models: **a** errors for sources in Tunnel A; **b** errors for sources in Tunnel B.

6.5 Discussion

The numerical simulations and field experiments presented in this study demonstrate that the proposed location method, which relies on an updated inverted heterogeneous velocity model, significantly enhances location accuracy. Nevertheless, several underlying assumptions, sources of error, and potential applications warrant further discussion. The key aspects are addressed as follows.

(1) Location error analysis

Errors affecting the proposed location method can be broadly attributed to two components: velocity inversion and source location.

In the velocity inversion stage, accurate estimation of the blasting initiation time t_0 is essential for P-wave traveltime tomography. The proposed approach determines t_0 by first inverting an S-wave velocity model through surface wave dispersion analysis, followed by back-propagation of S-waves from sensors to the source. This method relies on the principle that both P- and S-waves share the same trigger time, while noting that dispersion inversion itself does not require knowledge of t_0. Errors in t_0 estimation primarily arise from inaccuracies in the inverted S-wave velocity. The surface wave dispersion inversion is founded on a horizontal layer assumption (Park et al. 1998). Although the EDZ exhibits a layered velocity structure consistent with this assumption, its spatial extent may vary along the tunnel axis. In cases where the EDZ depth changes abruptly, the resulting t_0 estimation error may increase. Under complex geological conditions, direct measurement of t_0 using a trigger instrument is recommended to eliminate such errors entirely.

Another factor influencing inversion accuracy is the ray path density in P-wave tomography, which is constrained by the limited number of sources and sensors. Velocity updates are most effective in regions with dense ray coverage, whereas areas lacking ray paths remain unconstrained. To mitigate uneven updates, a smooth Gaussian filter may be applied. Further research into ray path density-based weighting functions could enhance tomographic inversion algorithms.

In the source location stage, accuracy is affected by both the ill-posed nature of the monitoring network and the simplification to two dimensions. Due to construction constraints, sensor layouts cannot fully enclose the monitoring area, and MS events often occur outside the sensor array. While the double-difference traveltime objective function combined with the simplex method improves convergence, the quasilinear sensor configuration remains sensitive to source locations aligned with the array axis, as evident in the location results for blasting sources A1 and B1 (Fig. 6.20). Moreover, the current inversion is limited to 2D due to the available acquisition geometry, which may introduce additional errors. Future deployment of 3D sensor networks would enable inversion of a three-dimensional heterogeneous velocity model and allow source location in a full 3D coordinate system.

(2) Identification of P- and S-wave first arrivals

Accurate picking of P- and S-wave first arrivals is critical for both traveltime tomography and source localization. As shown in the numerical and field waveforms

(Figs. 6.6 and 6.15), the P-wave first break is generally straightforward to identify. However, for near-field sensors, the overlap between P- and S-waves complicates reliable S-wave picking. Future studies may explore arrival identification based on instantaneous phase differences between P- and S-waves to reduce picking errors.

In this study, only one S-wave arrival—from the farthest offset sensor, where P- and S-waves are fully separated—is required to estimate t_0 (see Figs. 6.6a and 6.15c, g). For P-wave tomography, all available P-wave arrivals are used to ensure sufficient ray path coverage. The continuously updated heterogeneous P-wave velocity model offers higher resolution of geological structures compared to the layered S-wave model and captures velocity changes throughout excavation.

(3) Selection of inverted velocity models

Multiple inverted velocity models are generated at different stages of tunnel excavation. Numerical and field tests evaluated location accuracy using these models, with detailed results provided in Fig. 6.2. Although it might be expected that models incorporating more blasting sources would consistently yield better accuracy, the results indicate that the highest resolution is achieved in depth ranges corresponding to newly added blasting sources in the tomography. For instance, in the field case using Model 4 (black dots in Fig. 6.19), the locations of sources A7 and B7 are the most accurate. This improvement arises partly because deeper sources benefit from a more robust monitoring geometry and, more importantly, because tomographic updates preferentially adjust the velocity model along ray paths connecting new sources to sensors, thereby reducing location errors in adjacent areas. Therefore, it is advisable to select the inverted velocity model according to the specific monitoring region of interest.

(4) Potential application in linearly excavated tunnels

The numerical and field studies presented are based on two new tunnels intersecting an existing tunnel, where a well-developed EDZ significantly influences the velocity distribution. The proposed method demonstrates clear advantages in location accuracy and EDZ characterization for such intersecting tunnel configurations. Its applicability to conventional linear tunnel excavation is a subject of ongoing research. In linear tunnels, estimating t_0 becomes more challenging due to the greater distance between sensors and blasting sources, and a single inverted S-wave velocity profile may not adequately represent a layered model across the entire monitoring area. Direct measurement of t_0 using a trigger instrument at the source is a viable alternative. For P-wave tomography, a moving sensor array can provide superior acquisition geometry compared to the fixed configuration used here. Moreover, the proposed location algorithm can be readily extended to three dimensions by employing 3D FMM and simplex methods, provided that a 3D velocity model can be inverted using a suitably designed 3D monitoring network.

6.6 Conclusion

This chapter presents an MS source location method developed for deeply buried tunnels, which leverages an updated inverted heterogeneous velocity model. An improved perturbational dispersion inversion of surface waves and an adjoint-state traveltime tomography method utilizing blasting data are employed to reconstruct a heterogeneous P-wave velocity model. A double-difference simplex-based algorithm is introduced for MS source localization. The accuracy of the proposed method is validated through both numerical simulations and a field case study. The main conclusions are summarized as follows:

(1) The EDZ substantially alters wave propagation paths in deep tunnels. In accordance with Fermat's principle, waves emanating from an MS source travel through deeper, higher-velocity zones before bending back toward the sensors. Inverting the velocity structure of the EDZ is therefore a prerequisite for accurate MS location. Drilling and blasting source data provide a viable means for seismic exploration. The dispersion of surface waves propagating along the tunnel sidewall enables inversion of the S-wave velocity within the EDZ. An improved perturbational dispersion inversion method, combining a modified thin-layer method for forward simulation with perturbational inversion, is developed to estimate both the S-wave velocity profile and the depth of the EDZ.
(2) Heterogeneous P-wave velocity models can be obtained through first-arrival traveltime tomography of blasting sources. The initiation time of each blasting source is estimated by backpropagating S-waves using the FMM applied to the inverted S-wave model. An adjoint-state first-arrival tomography approach is adopted in this study. The FMM solves the eikonal equation to compute forward traveltimes, while the adjoint-state method computes the gradient of the misfit function without requiring explicit Fréchet derivatives. As new blasting source data become available, the velocity model is iteratively updated.
(3) An objective function for MS source location is formulated based on the L2 norm of the double-difference traveltimes. The FMM is used to forward-simulate traveltimes within the inverted heterogeneous P-wave model, from which the double-difference traveltimes are derived. This formulation eliminates the need for the origin time of the MS source and enhances the stability of the location procedure. The simplex method is employed as a nonlinear optimization technique to minimize the objective function and determine the optimal source location.
(4) Numerical simulations based on a realistic tunnel model are conducted to evaluate the location accuracy of the proposed method. Wavefield snapshots provide visualization of the blasting source wave propagation, including P-waves, S-waves, and surface waves traveling through the surrounding rock along the tunnel sidewall. Surface wave dispersion analysis successfully inverts a layered S-wave model, enabling accurate estimation of t_0 via the FMM. First-arrival traveltime tomography yields detailed heterogeneous P-wave velocity models. The location accuracy is tested using 64 simulated MS sources distributed

around the tunnel face. The results demonstrate that the location based on the inverted heterogeneous velocity model significantly outperforms that based on homogeneous velocity assumptions. Furthermore, the location accuracy generally improves as more blasting sources are incorporated into the tomographic inversion.

(5) A field experiment is carried out at two newly excavated tunnels with a burial depth of 2400 m in the Jinping Underground Laboratory Phase II, one of the deepest underground laboratories worldwide. Heterogeneous P-wave velocity models of the monitored area are continuously inverted throughout tunnel excavation using blasting source data. Blasting events with precisely known locations are used to validate the proposed location method. The results indicate that most location errors are within 5 m when inverted velocity models are employed. With appropriate selection of the updated inverted model, location errors can be reduced to less than 2 m.

References

Castellanos F, Van der Baan M (2015) Dynamic triggering of microseismicity in a mine setting. Geophys J Int 202(2):728–737

Chen BR, Li T, Zhu XH et al (2021) Microseismic source location method based on a velocity model database and statistical analysis. Arab J Geosci 14(19):1–16

Collins DS, Toya Y, Pinnock I, et al. 2014 3D velocity model with complex geology and voids for microseismic location and mechanism. In: Deep mining 2014: proceedings of the seventh international conference on deep and high stress mining. Australian Centre for Geomechanics, pp 681–688

Dong L, Li X, Tang L et al (2011) Mathematical functions and parameters for microseismic source location without pre-measuring speed. Chin J Rock Mech Eng 30(10):2057–2067

Dong L, Shu W, Han G et al (2017a) A multi-step source localization method with narrowing velocity interval of cyber-physical systems in buildings. IEEE Access 5:20207–20219

Dong L, Shu W, Li X et al (2017b) Three dimensional comprehensive analytical solutions for locating sources of sensor networks in unknown velocity mining system. IEEE Access 5:11337–11351

Dong L, Hu Q, Tong X et al (2020) Velocity-free MS/AE source location method for three-dimensional hole-containing structures. Engineering 6(7):827–834

Feng GL, Feng XT, Chen BR et al (2015) Sectional velocity model for microseismic source location in tunnels. Tunnell Underground Space Technol 45:73–83

Haney MM, Tsai VC (2017) Perturbational and nonperturbational inversion of Rayleigh-wave velocities. Inversion of Rayleigh-wave velocities. Geophysics 82(3):F15–F28

Hey F, Chenw T, Chen F (2006) Normal modes computation by generalized reflection-transmission coefficients method in planar layered half space. Chin J Geophys 49(4):965–973

Jiang R, Dai F, Liu Y et al (2021) Fast marching method for microseismic source location in cavern-containing rockmass: performance analysis and engineering application. Engineering 7(7):1023–1034

Kausel E, Roësset JM (1981) Stiffness matrices for layered soils. Bull Seismol Soc Am 71(6):1743–1761

Li N, Wang E, Ge M et al (2014) A nonlinear microseismic source location method based on Simplex method and its residual analysis. Arab J Geosci 7(11):4477–4486

Liu L, Wang Y (2021) Perturbational inversion of fundamental mode Rayleigh wave dispersion curve. In: SEG/AAPG/SEPM First international meeting for applied geoscience & energy. OnePetro

Liu L, Li S, Xiao Y et al (2023) An inverted heterogeneous velocity model for microseismic source location in deep buried tunnels. Rock Mech Rock Eng 56(7):4855–4880

Moler CB, Stewart GW (1973) An algorithm for generalized matrix eigenvalue problems. SIAM J Numer Anal 10(2):241–256

Nelder JA, Mead R (1965) A simplex method for function minimization. Comput J 7(4):308–313

Park CB, Miller RD, Xia J (1998) Imaging dispersion curves of surface waves on multi-channel record. In: SEG technical program expanded abstracts 1998. Society of Exploration Geophysicists, pp 1377–1380

Taillandier C, Noble M, Chauris H et al (2009) First-arrival traveltime tomography based on the adjoint-state method. Geophysics 74(6):WCB1-WCB10

Xiao Y, Chen S, Liu L et al (2026) A stratified double-difference dynamic tomography of 3D velocity field for deep-buried large underground cavern: Y. Xiao et al. Rock Mech Rock Eng:1–24

Xu Y, Xia J, Miller RD (2007) Numerical investigation of implementation of air-earth boundary by acoustic-elastic boundary approach. Geophysics 72(5):SM147-SM153

Zhao H (2005) A fast sweeping method for eikonal equations. Mathematics of computation 74(250):603–627

Zheng M, Li S, Zhao H et al (2021) Probabilistic analysis of tunnel displacements based on correlative recognition of rock mass parameters. Geosci Front 12(4):101136

Chapter 7
Analyzing the Stability of Rock Surrounding Deep Cross-Tunnels Using a Dynamic Velocity Field

7.1 Prologue

Deeply buried engineering projects often involve intersecting tunnel systems, where new excavations are constructed in proximity to existing excavations to meet specific functional requirements (Xiao et al. 2024, 2025). Such interactions can compromise the stability of the surrounding rock, potentially leading to significant fracturing and deformation (Verruijt and Booker 1998). Consequently, assessing the stability of the rock mass in cross-tunnel configurations is essential for safeguarding both personnel and equipment. The excavation-damaged zone (EDZ) serves as a key quantitative indicator for evaluating the effects of excavation and subsequent stress redistribution in the surrounding rock (Song et al. 2021; Chen et al. 2023). Excessive damage within the EDZ can substantially undermine structural stability, making its characterization a critical component of stability analysis (Chen and Wang 2022). The overlapping of old and new tunnels introduces complex stress interactions, exacerbated by blasting and support activities, which further complicate the evolution of the EDZ (Zhu et al. 2014).

Monitoring EDZ development using drilling or blasting data has proven to be a viable approach (Liu et al. 2023; Ma et al. 2023). Specifically, dynamic monitoring of the EDZ can be achieved through inversion of the seismic velocity field (Ozaki 2023). However, a major challenge in traveltime tomography lies in the imprecise timing information associated with drilling or blasting events, as accurate timing is essential for reliable traveltime calculations (Hu et al. 2021). Previous research has established that Rayleigh waves are sensitive to EDZ characteristics (Leparoux et al. 2012) and can be effectively used to invert a layered S-wave velocity model (Czarny et al. 2021). This provides a practical means to estimate the S-wave structure of the EDZ, from which the initiation time of a drilling or blasting signal can be derived via back-propagation of rays within the layered model.

First-arrival traveltime tomography is widely recognized as an effective technique for inverting velocity fields (Taillandier et al. 2009). Various inversion strategies

© Tongji University Press Co., Ltd 2026

Y. Xiao et al., *Near-Field Microseismic Data Processing in Deep Tunnels*,
https://doi.org/10.1007/978-981-92-3208-6_7

have been developed based on different forward-modeling approaches, including ray-tracing methods using shortest-path or curved-ray algorithms, as well as techniques that solve the eikonal equation via the adjoint-state method (Wang and Park 2001). The use of the fast marching method (FMM) or fast sweeping method to solve the eikonal equation considerably reduces computational complexity, thereby enhancing inversion speed. The velocity field of a rock mass is closely related to its mechanical properties (Zhu et al. 2021). Continuous updating of the velocity field enables real-time estimation of these properties, which can subsequently be incorporated into numerical simulations to predict rock mass stability. Numerical modeling plays a central role in stability analysis and hazard assessment in rock engineering. Commonly employed methods include the finite element method, finite difference method, discrete element method, particle flow code, finite discrete element method, and continuum–discontinuum finite element method (Müller 1991; Feng et al. 2006; Hadjigeorgiou et al. 2009; Lin et al. 2023).

Deep rock engineering is distinguished by high in situ stresses, complex spatial configurations, significant postexcavation stress path alterations, and a prevalence of tension-induced fracturing (Pan et al. 2006). To address these challenges, researchers have developed three-dimensional numerical analysis tools and mechanical models capable of capturing the geological conditions and deformation behaviors characteristic of deep rock masses (Hajiabdolmajid et al. 2002; Ma et al. 2022). Appropriate numerical indicators are required to evaluate rock mass stability. Widely used computational indicators include the yield approach index, failure approach index, brittleness evaluation index, equivalent plastic strain index, energy release rate, and EDZ assessment (Feng et al. 2021). More recently, the degree of rock fracture and the local energy release rate (LERR) have been introduced to characterize tension-dominated failure mechanisms in deep rock, providing insight into rupture processes, their locations and severity, and the local energy released during excavation (Sakurai 2017).

The mechanical parameters selected to represent a rock mass profoundly influence numerical simulation outcomes. These parameters are typically determined through laboratory tests, inverse analysis, or direct in situ measurements (Guo et al. 2017). However, laboratory results often cannot be directly applied, and in situ measurements, particularly those aimed at determining the cohesion or deformation modulus, are frequently costly and technically demanding (Feng et al. 2018). Although recent advances allow deformation modulus estimation via field displacement measurements combined with inverse analysis (Park et al. 1998), the accuracy remains constrained by the limited precision of multipoint displacement meters or tunnel convergence meters. Moreover, such methods involve significant time delays, hindering real-time assessment. Therefore, there is an urgent need for numerical simulation approaches that can accurately reflect the real-time evolution of mechanical parameters, enabling precise stability analyses in deep rock engineering.

In this study, we employ traveltime tomography to invert a heterogeneous P-wave velocity field using drilling and blasting data. The inverted velocity field is then converted into rock material parameters, allowing continuous updating of these parameter fields. These inputs are subsequently integrated into Cellular Automation

Software for engineering Rockmass fracturing processes (CASRock) to perform numerical simulations and stability analyses. Rock damage is evaluated using two key indicators: equivalent plastic shear strain (eps) and local energy release rate (LERR). This work aims to establish a foundation for ensuring the safety of personnel and operations in deep tunnel environments.

7.2 Methodology

7.2.1 Overall Workflow

The stability analysis method proposed in this study follows a two-step workflow, as illustrated in Fig. 7.1. The first step involves the inversion of the P-wave velocity field, while the second step focuses on the stability analysis of the surrounding rock. Each step is described in detail below.

Step 1: Inversion of the P-wave Velocity Field

Prior to inverting the P-wave velocity field, it is essential to accurately determine the initiation time t_0 of the blasting event. A dispersion curve is first extracted from the blast wave signals using the phase-shift method (Park et al. 1998). Through an iterative inversion process, the S-wave velocity field is reconstructed by matching the dispersion curve derived from the inverted model with that of the actual S-wave velocity field. The eikonal equation is then solved for the inverted S-wave velocity model using FMM to compute the traveltimes from the source to each sensor. The initiation time t_0 is subsequently obtained as the difference between the inverted traveltime and the observed S-wave arrival time.

Following each blasting event, adjoint-state traveltime tomography is performed to reconstruct the P-wave velocity field. The absolute P-wave traveltime is calculated as the difference between t_0 and the P-wave arrival time. An initial P-wave velocity model for tomography is estimated based on the slope of the P-wave event axis in the wavefield records and the previously obtained layered S-wave model. The inversion process iteratively updates the velocity model until the misfit between the simulated and observed absolute traveltimes is minimized. The resulting velocity field represents the region between the source and the sensor array. This procedure is repeated after each drilling or blasting cycle, enabling continuous updating of the P-wave velocity field as excavation progresses.

Step 2: Stability Analysis of the Surrounding Rock

For rock materials, seismic wave velocity is generally correlated with key mechanical properties, including Poisson's ratio, elastic modulus, and density. In this study, Poisson's ratio and density are assumed to remain constant during excavation to facilitate approximate calculations. Based on this assumption, the elastic modulus of the surrounding rock is derived from the inverted P-wave velocity field. The resulting

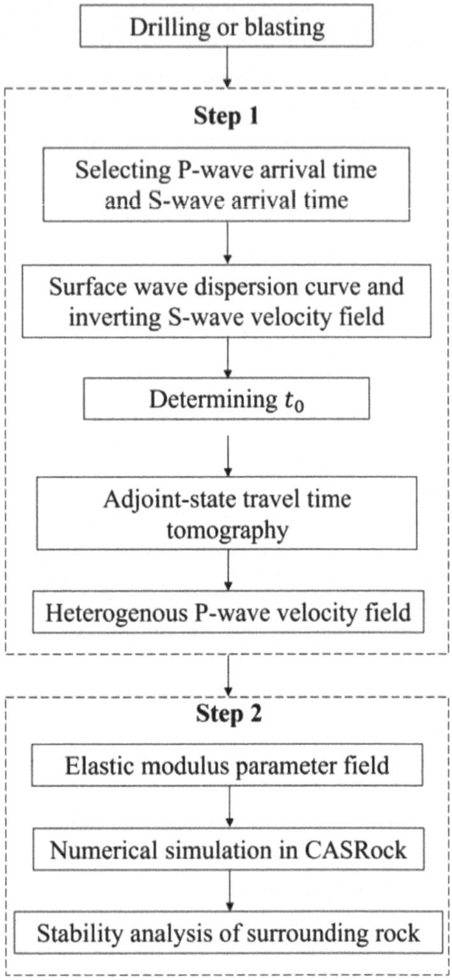

Fig. 7.1 Overall workflow involved in the new stability analysis method

spatially distributed elastic modulus is then imported into CASRock V1.0 software for numerical simulation. The extent and severity of rock damage are evaluated using two indicators, eps and LERR, both of which are extracted from the CASRock simulation results.

7.2.2 Adjoint-State Traveltime Tomography

Please refer to Sect. 6.4.2.3.

7.2.3 *Stability Analysis with Dynamically Updated Mechanical Properties*

7.2.3.1 Updating the Mechanical Properties

Following the procedure in Sect. 7.2.2, the wave velocity field at a specific excavation stage is obtained. For real-time updating of mechanical parameters, this velocity field must be mapped to a corresponding mechanical parameter field. In the case of rock, the wave velocity is commonly calculated using:

$$v = \sqrt{\frac{E(1-\mu)}{\rho(1+\mu)(1-2\mu)}} \tag{7.1}$$

E, μ, and ρ correspond to the elastic modulus, Poisson's ratio, and density of the rock, respectively. The elastic modulus is strongly dependent on the material's loading and unloading history during excavation, an effect commonly termed damage. The fundamental definition of damage can be expressed as:

$$D = \frac{E_c}{E_0} \tag{7.2}$$

where E_c and E_0 are the current and initial elastic modulus, respectively. Combining with Eq. 7.1 leads to the relationship between these moduli and their associated wave velocities v_c and v_0:

$$\frac{E_c}{E_0} = \frac{v_c^2}{v_0^2} \tag{7.3}$$

where v_c represents the current wave velocity and v_0 represents its initial wave velocity.

It is worth noting that Eq. 7.1 contains three material parameters. To facilitate computation, Poisson's ratio and density are taken as constants during excavation and unloading. With this simplification, the elastic modulus at a given point can be calculated efficiently.

7.2.3.2 Simulation Platform

CASRock is a numerical modeling platform designed for geotechnical analysis of engineering rock masses (Feng et al. 2006). It supports engineering design, safety factor prediction, research and testing, and failure back-analysis. The software models the rock mass as an assembly of heterogeneous deformable cells that interact with each other. It simulates local mechanical responses—such as movement and deformation—for each individual cell, accounting for effects from neighboring cells, including external, elastoplastic, and inertial forces; chemical or biological influences; thermal and seepage processes; cracking; invasion; and friction. A cell's state

Table 7.1 Main parameters

Parameters						
Initial stress conditions	Elastic modulus	Poisson'sratio	Cohesion	Inter friction angle	Tensile strength	Plastic strain

progresses continuously from continuum to discontinuum, encompassing elastic, plastic, and failure stages, which is quantified by an internal variable. Interactions among cells evolve in a self-organizing manner from the local level outward, with the evolution governed by the state of the local cell. Global equilibrium is attained after local calculations are completed and their effects are propagated. Comprehensive descriptions of CASRock's principles are available in Feng et al. (2021).

In this study, CASRock was employed to assess the stability of the surrounding rock. The Mohr–Coulomb strength criterion with tensile cutoff (He et al. 2019) and the cohesion weakening and friction strengthening (CWFS) model (Walton 2019) were adopted for the rock material. This approach typically requires nine input parameters, as shown in Table 7.1. Notably, the elastic modulus is updated in real time during calculations based on the wave velocity field described in Sect. 7.2.2. The main parameters required for CASRock simulations are summarized in the following list.

7.2.3.3 Stability Analysis

In this study, eps and LERR are employed to evaluate the stability of the surrounding rock (Feng et al. 2021). The eps indicator captures key characteristics of rock damage, including the size of the plastic zone, damage depth, spatial extent, and severity. Its scientific validity and practical applicability have been well established through extensive use in engineering stability analyses (Hajiabdolmajid et al. 2002). The LERR indicator, on the other hand, quantifies the amount of energy released during the damage process and serves as a complementary measure of damage severity. It has been widely adopted in studies investigating brittle failure phenomena, such as rockbursts (Jiang et al. 2010).

7.3 Case Study

The proposed method is employed here to analyze the stability of the surrounding rock in the Jinping tunnel, where a new tunnel (tunnel B) is excavated using drilling and blasting. Stability assessments are conducted after each blasting event by feeding the dynamically updated elastic modulus, derived from P-wave velocity inversion, into CASRock simulations. The input parameters are detailed in Table 7.1; the elastic modulus is updated in real time, while other parameters are estimated from previous

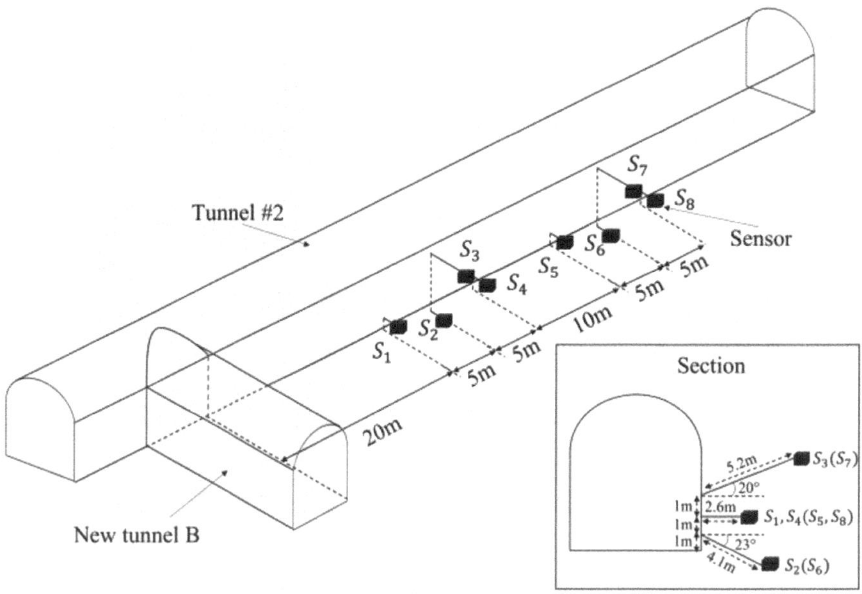

Fig. 7.2 MS sensor layout in tunnel B

Jinping studies (Feng et al. 2020; Jiang et al. 2016; Cui et al. 2019). Section 6.4.1 provides further background. The MS sensor layout and labeling are depicted in Fig. 7.2.

7.3.1 Tomography Results

The waveforms recorded by the eight sensors during the first blasting event in the excavation of new tunnel B are presented in Fig. 7.3a. Multiple explosive charges were detonated within a single blasting sequence, and the resulting wave trains were clearly distinguishable, with minimal interference between successive events. The initial blast wave was selected for inversion of the P-wave velocity field. An enlarged view of this first explosion waveform is shown in Fig. 7.3b, where the P-wave arrivals are readily identifiable through manual picking. Among the eight sensors, S_1 to S_4 are positioned closer to the blasting source than S_5 to S_8, rendering the S-wave arrivals more discernible at the more distant sensors. Figure 7.3c illustrates the sequential propagation of the wave from the source to sensors S_1, S_4, S_5, and S_8.

As the source-to-sensor distance increases, the separation between P- and S-wave arrivals becomes progressively more pronounced. Furthermore, the waveforms and arrival times recorded by sensor pairs S_2 and S_3, as well as S_6 and S_7, exhibit strong similarity. Considering this together with the sensor layout illustrated in Fig. 7.2,

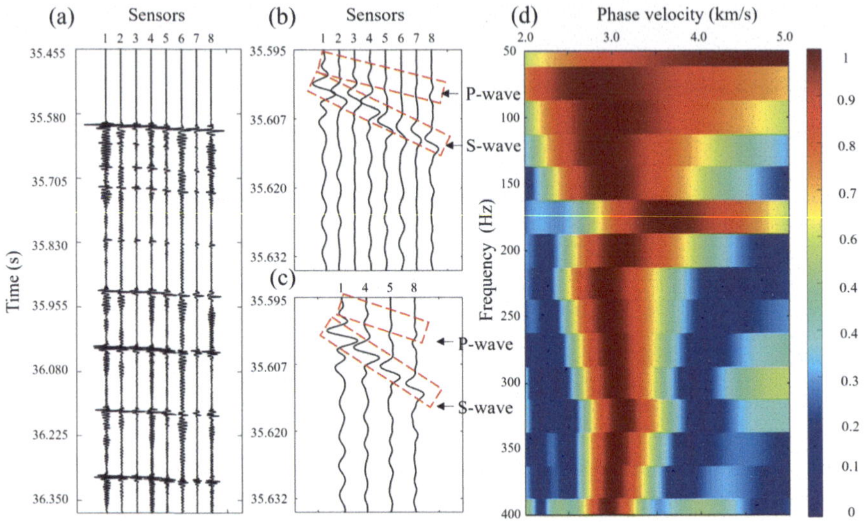

Fig. 7.3 Processing of blasting signals from tunnel B: **a** waveform of the first blast event; **b** Time slice corresponding to the first pulse in (**a**); **c** aligned waveforms of channels S_1, S_4, S_5, and S_8 for the first pulse; **d** dispersion curve extracted from the tunnel B blasting data

where S_2/S_6 and S_3/S_7 are positioned at different elevations relative to the others, it can be inferred that the velocity field is well suited to 2D simulation.

Figure 7.3d presents the dispersion curve extracted from the blasting signal, which illustrates the variation in the surface wave phase velocity with frequency. Lower phase velocities are observed at higher frequencies, whereas higher velocities are concentrated in the frequency range of 50–175 Hz. Based on this dispersion curve, the layered S-wave velocity model is inverted using an improved perturbational inversion method (Liu and Wang 2021).

The initial layered S-wave velocity model assumes a monotonic increase in velocity with depth, as depicted in Fig. 7.4b. In contrast, the final inverted model exhibits a nonlinear velocity variation with depth. As shown in Fig. 7.4a, the dispersion curve calculated from the inversion results closely matches the observed dispersion curve, confirming that the inverted layered S-wave velocity model is both reasonable and accurate.

The FMM is subsequently employed to solve the eikonal equation based on the inverted layered S-wave velocity model, yielding the traveltimes of S-waves propagating from the blasting source to each of the eight sensors. By integrating these computed traveltimes with the manually picked S-wave arrival times, the blasting initiation time t_0 is determined to be 23:19:51.64450 on September 1, 2021. This value is then used to calculate the absolute P-wave traveltimes and to perform P-wave velocity field tomography via the adjoint-state method. The resulting imaging outcomes are presented in Fig. 7.5.

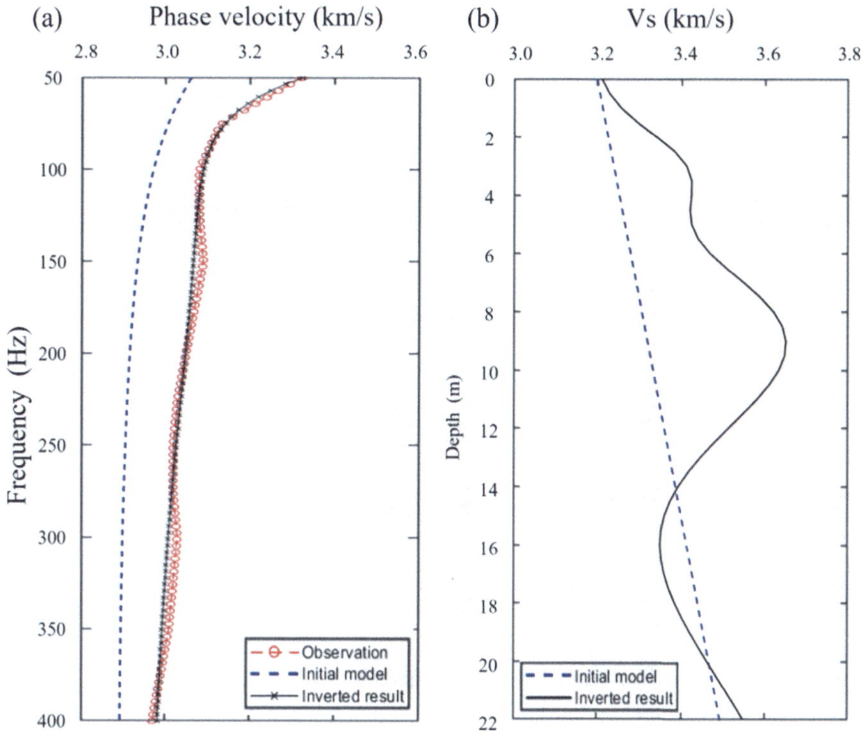

Fig. 7.4 Results obtained from dispersion curve inversion: **a** comparison of dispersion curves; **b** comparison of S-wave velocity models

The initial velocity field is depicted in Fig. 7.5g, while Fig. 7.5a–f illustrate the evolution of the velocity field between the source and the sensor array as excavation progressed. The inversion results reveal a persistent high-velocity region (highlighted by the white dashed box in Fig. 7.5d) consistently observed across Figs. 7.5a–d, indicating relatively higher velocities at depths of 4–10 m. In contrast, the velocity field within the 15–20 m depth range remains largely unchanged during the early stages, as ray paths from shallower sources do not penetrate this zone. However, as source depths increase, rays gradually traverse these deeper regions, leading to velocity updates evident in Fig. 7.5e, f.

A distinct low-velocity zone is identified in the black dashed box in Fig. 7.5f, where velocities are markedly lower than those in the 4–10 m depth interval. The inverted velocity fields thus enable the identification of EDZs at varying depths. In this case, the results suggest that the rock mass at depths of 15–20 m exhibits relatively poor integrity.

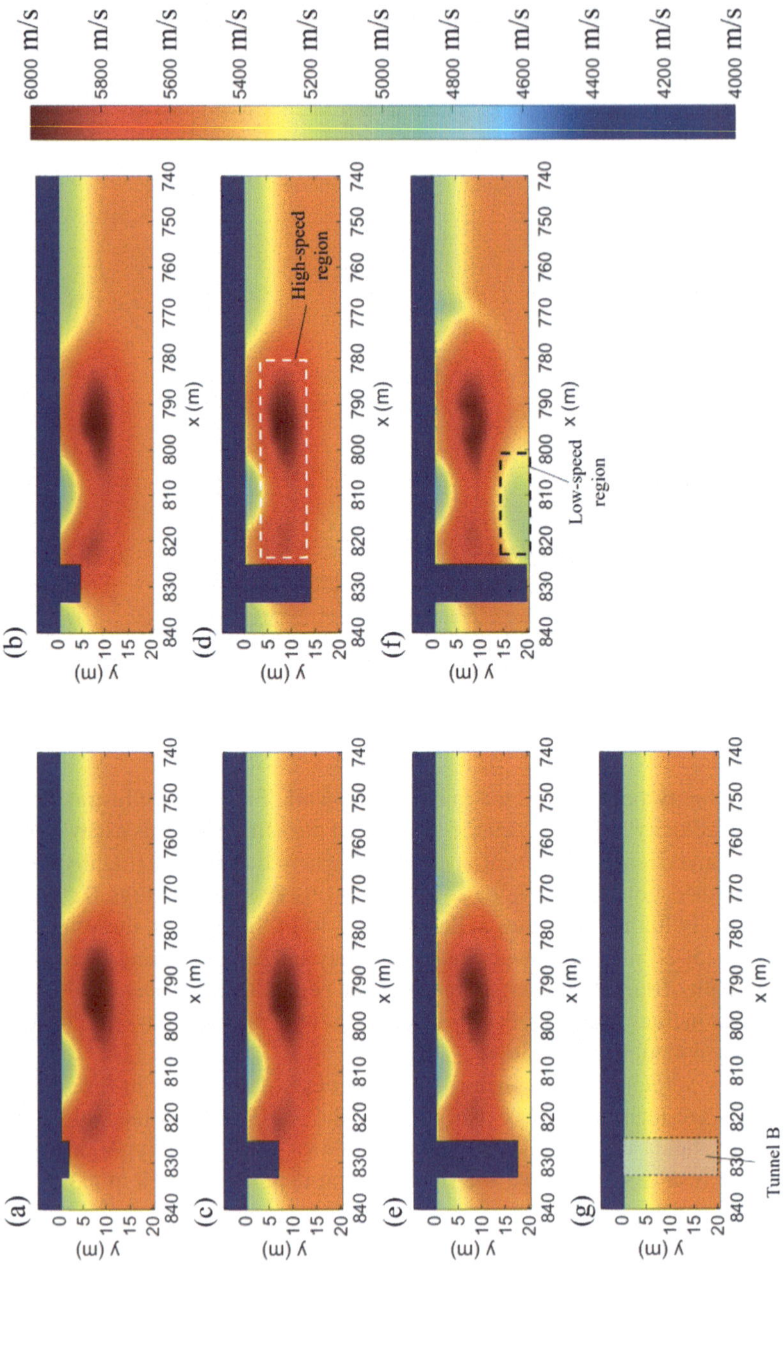

Fig. 7.5 P-wave velocity models reconstructed by traveltime tomography during excavation: **a–f** inversion results using the first to sixth blasting sources sequentially; **g** initial velocity model

7.3.2 Stability Analysis Results

7.3.2.1 Simulation Setup

Figure 7.6 presents the numerical model configuration adopted in this study. As illustrated in Fig. 7.6a, a rectangular domain of dimensions 100 m × 100 m × 200 m is established. The longer dimension (200 m) aligns with the axis of auxiliary tunnel #2, which traverses the entire model. The primary focus of this investigation is new tunnel B, situated near the center of the model. The model dimensions are chosen to ensure that the boundaries are positioned at distances equivalent to three to five times the diameters of both auxiliary tunnel #2 and tunnel B.

During the loading simulation, displacement constraints are imposed on all six boundaries of the rectangular model. Specifically, the left and right faces are fixed in the x-direction, the front and back faces are fixed in the y-direction, and the top and bottom faces are fixed in the z-direction. Figure 7.6b depicts the mesh discretization employed for tunnels #2 and B. A tetrahedral meshing scheme is adopted throughout the model. The grid resolution is set to 0.5 m for tunnel #2, 0.1 m for tunnel B, and 2 m for the remaining regions. In total, the model is discretized into 101,706 elements of varying sizes. Following the completion of the auxiliary tunnel excavation, the

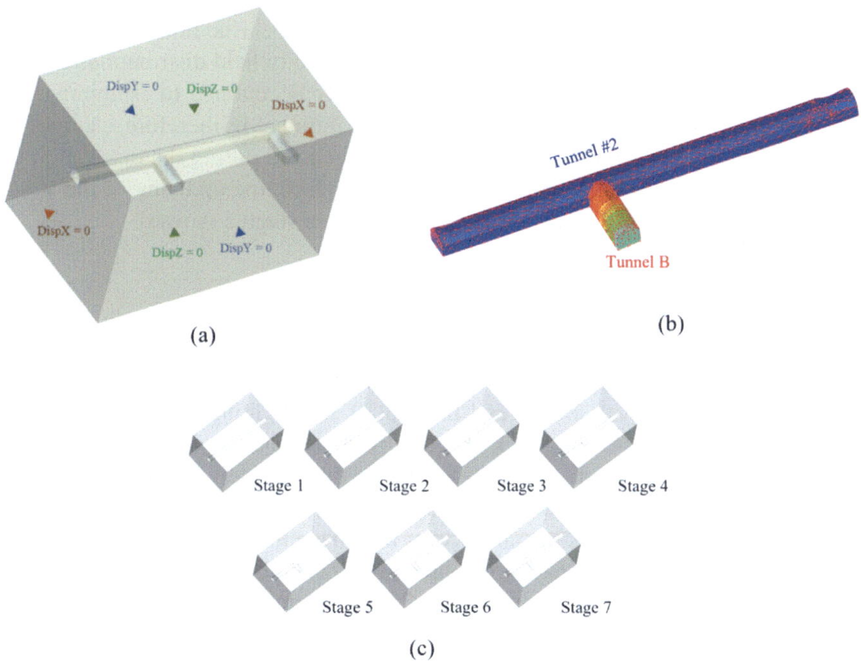

Fig. 7.6 Simulation framework details: **a** model geometry and boundary constraints; **b** grid employed for tunnel modeling; **c** excavation stages

Table 7.2 Initial stress conditions

σ_x (MPa)	σ_y (MPa)	σ_z (MPa)	τ_{xy} (MPa)	τ_{yz} (MPa)	τ_{zx} (MPa)
47.56	54.55	62.19	4.52	2.35	15.14

Table 7.3 Main mechanical parameters

Parameter	Initial	Peak stress	Residual
Elastic modulus (GPa)	29.2	Updated	Updated
Poisson's ratio	0.23	0.23	0.23
Cohesion (MPa)	35.0	25.9	7.8
Internal friction angle (°)	30.0	37.0	51.0
Tensile strength (MPa)	1.5	1.5	1.5
Plastic strain (‰)	0.0	2.0	6.0

excavation of tunnel B is simulated according to the actual construction schedule. The specific excavation stages are illustrated in Fig. 7.6c.

The damage criterion and mechanical model described in Sect. 7.2.3.2 are employed in conjunction with the parameter values listed in Tables 7.2 and 7.3. The updating of the elastic modulus commences at stage 2 of the excavation sequence (Fig. 7.6c) and continues throughout the process following the algorithm presented in Sect. 7.2.3.1. Figure 7.7 illustrates the evolution of the elastic modulus at various excavation stages, with trends consistent with the velocity field distributions shown in Fig. 7.5. It should be noted that the present study is confined to examining the influence of auxiliary tunnel #2 on the excavation of tunnel B; therefore, changes in the wave velocity field and elastic modulus that occurred during the excavation of tunnel #2 itself are not taken into account. Furthermore, the updating of the wave velocity field discussed in Sect. 7.2.2 also begins from stage 2 onward.

7.3.2.2 Temporal Failure Characteristics

Figure 7.8 presents the eps contours on the *yz*-, *xz*-, and *xy*-plane slices at various stages of tunnel B excavation (stages 1 through 7), highlighting the development of plastic zones in the surrounding rock. It is observed that both the extent of the plastic zone and the associated damage intensity increase progressively during the first four excavation stages, with minimal changes thereafter from stage 5 onward. This indicates that the majority of rock damage occurs within the initial four stages, assuming no support intervention. The rate of damage growth is highest between stages 1 and 2, decreases progressively through stages 2–3 and 3–4, and becomes negligible from stage 5 to 7. These results suggest a gradual deceleration in damage development as excavation advances. Consequently, the most significant damage, both in extent and severity, takes place within the first 4.5 m of excavation, corresponding to the period prior to stage 3.

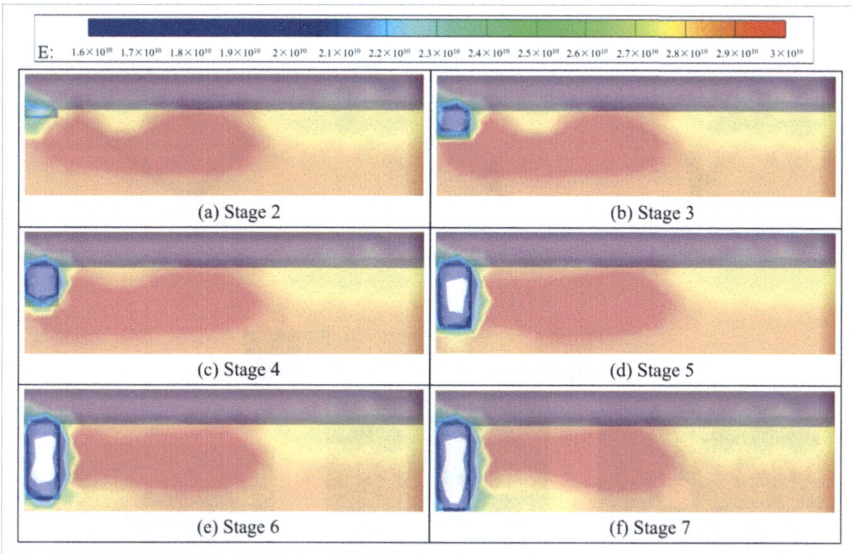

Fig. 7.7 Progressive update of the elastic modulus throughout excavation, showing results for stages **a** 2, **b** 3, **c** 4, **d** 5, **e** 6, and **f** 7 (stage numbering as in Fig. 7.6c)

7.3.2.3 Spatial Failure Characteristics

Figure 7.9 illustrates the distributions of eps and LERR around tunnel B at various distances from the auxiliary tunnel following the completion of excavation stage 7. The eps results, shown in the second column of Fig. 7.9, exhibit a clear spatial trend: the closer the section is to tunnel #2, the larger the plastic zone and the more severe the damage. At a distance of 0 m from tunnel #2 (Fig. 7.9b0), the damage is most extensive, with eps values exceeding 0.03 adjacent to the sidewalls. For distances ranging from 4 to 18 m, the maximum eps values fall between approximately 0.012 and 0.018, while at 16 m, the Epstn value is only approximately 0.006.

A similar pattern is observed in the LERR results: as the distance from tunnel #2 increases, both the spatial extent and magnitude of energy release decrease. It is worth noting that in Fig. 7.9b18, c18, the left sidewall exhibits slightly higher levels of damage and energy release, which can be attributed to its proximity to the excavated working face. Nevertheless, this localized effect does not alter the overarching trend that damage severity and energy release diminish with increasing distance from the auxiliary tunnel.

In this study, a threshold value of 0.006 for eps is adopted, as specified in Table 7.3. Rock material is considered to have undergone complete failure when eps exceeds this threshold. The term "yield depth" is introduced to denote the maximum distance from the tunnel sidewall at which eps values surpass 0.006. Quantitative analysis based on this definition enables a comparative evaluation of yield depth and maximum energy release rate, as presented in Fig. 7.10.

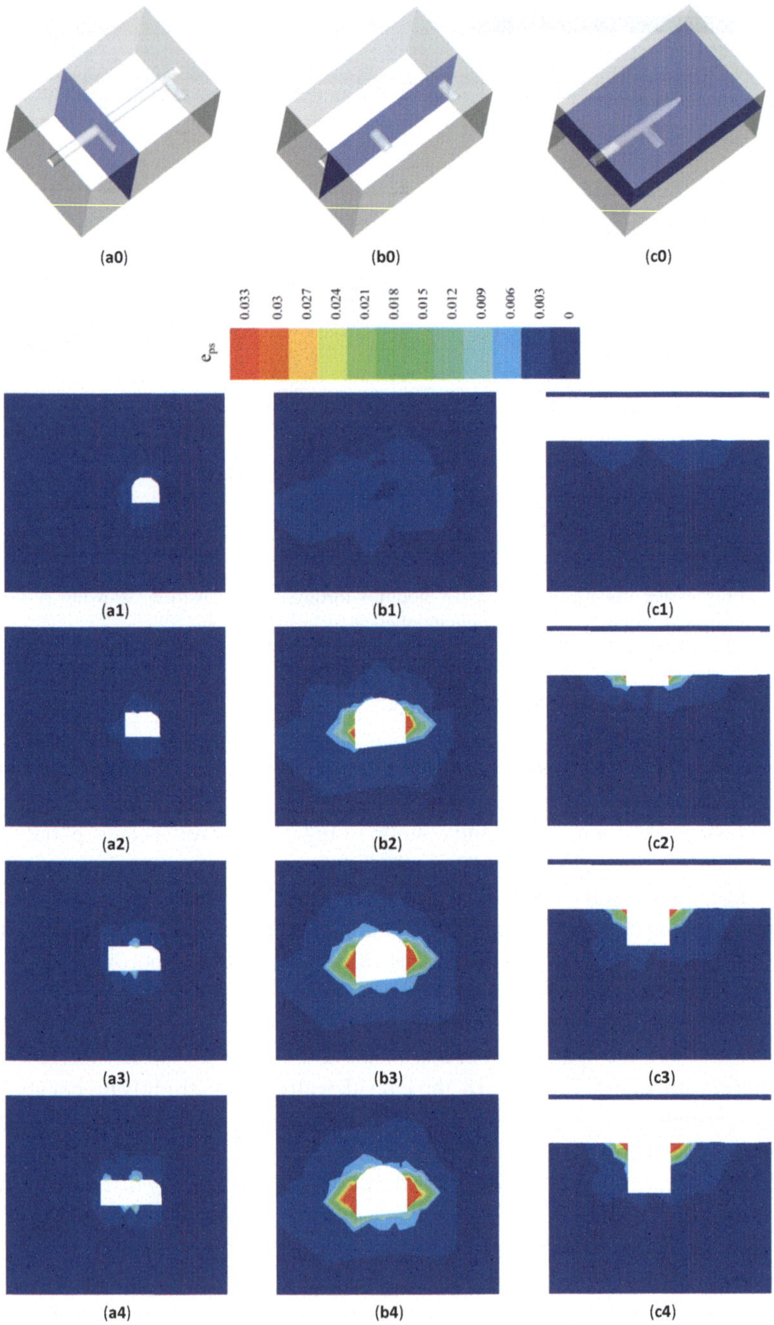

Fig. 7.8 Progressive change in eps contours during tunnel B excavation. Columns from left to right display slices in the *YZ*-, *XZ*-, and *XY*-planes (positions given in the top row). The subscript n ($1 \leq n \leq 7$) denotes the excavation stage

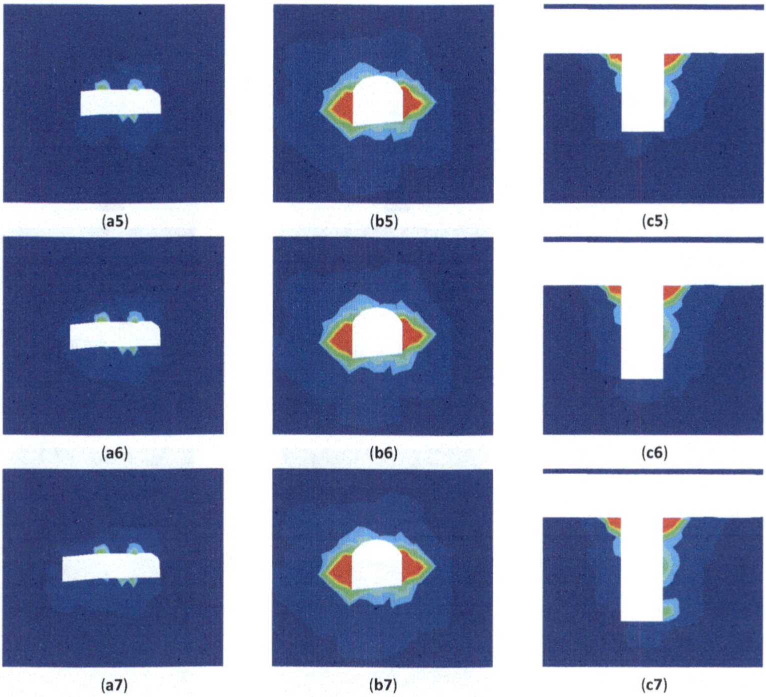

Fig. 7.8 (continued)

Figure 7.10a illustrates that both the yield depth and energy release decrease progressively with increasing distance from auxiliary tunnel #2, a trend consistent with the contour distributions shown in Fig. 7.9. The variation in yield depth across different tunnel sections is further depicted in Fig. 7.10b, which provides a cross-sectional view of its spatial distribution. This distribution map reveals that the largest yield depths are concentrated near the left and right sidewalls, reaching up to approximately 5 m, whereas the yield depths at the crown and invert are considerably smaller.

A more detailed examination of Fig. 7.10a indicates that the yield depth exhibits a fluctuating pattern: it peaks within the 0–2 m interval, declines over the 4–6 m range, and rises again in the 8–10 m zone, forming a high–low–high–low alternation. Similarly, the maximum energy release rate is greatest at 0 m and displays secondary local peaks at distances of 4 and 10 m. These observations may be attributed to the presence of a zone characterized by elevated wave velocity or elastic modulus within the 4–10 m range. In the following section, the influence of real-time updates to the elastic modulus on the evolution of damage zones and energy release patterns is examined in greater detail, with the aim of elucidating these intriguing phenomena.

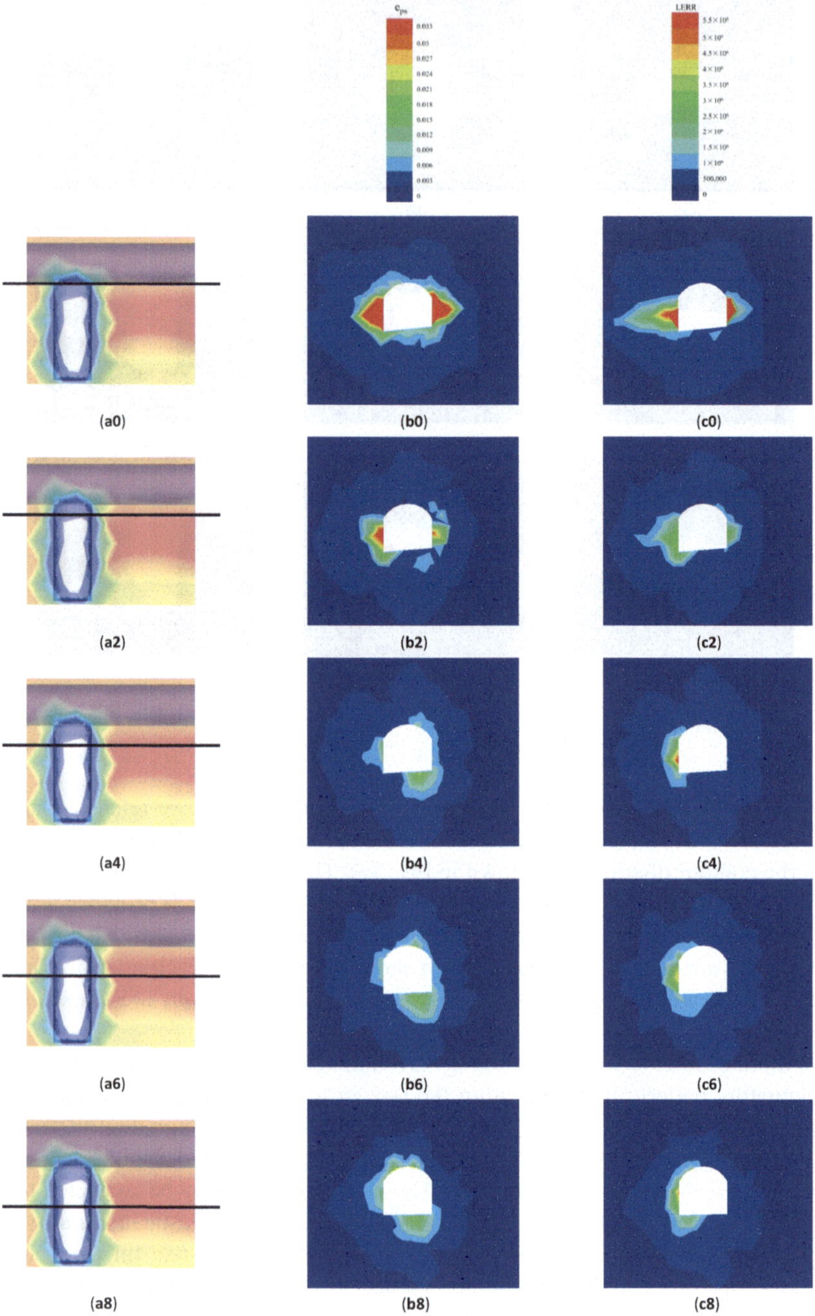

Fig. 7.9 Spatial variation in eps and LERR. **a** Plan view with the black line representing the section location. **b** eps contours on the section. **c** LERR contours on the section. The labels 0, 2, …, 18 correspond to distances from tunnel #2 in meters

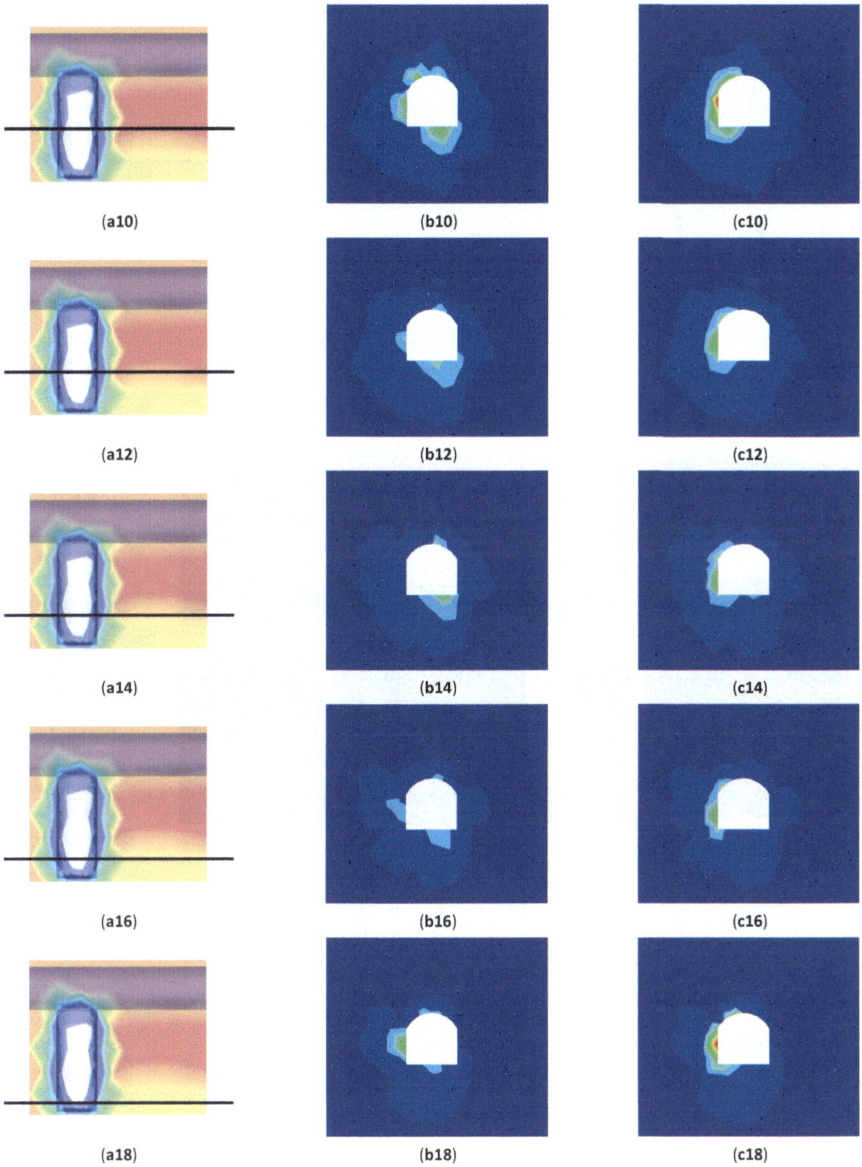

Fig. 7.9 (continued)

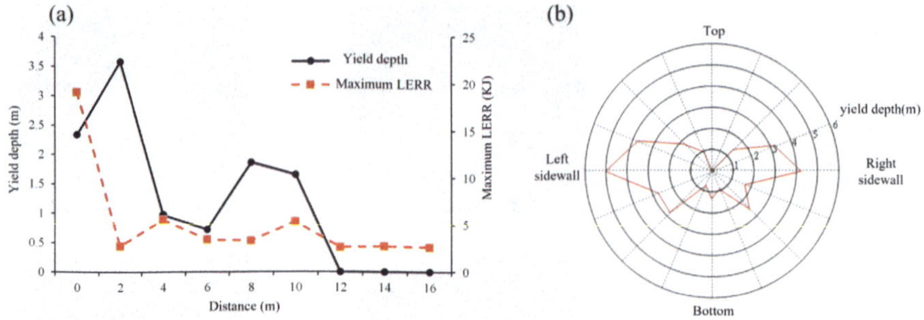

Fig. 7.10 Relationship between rock damage indicators and distance from tunnel #2. **a** Yield depth and maximum energy release rate as functions of distance; **b** cross-sectional view of the yield depth distribution in tunnel B

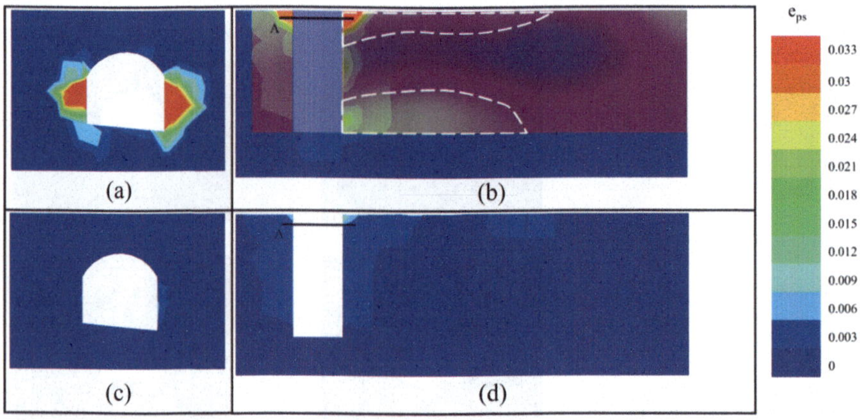

Fig. 7.11 eps results from the new method (top row) and conventional method (bottom row). **a** Section along line A in (**b**) with real-time elastic modulus updating; **b** plan view. **c** Section and **d** plan view for the constant-modulus simulation

7.4 Discussion

(1) Comparison with the Traditional Stability Analysis Method

A comparison is presented between the conventional stability analysis method and the approach proposed in this study, as illustrated in Fig. 7.11. The contour plots of the equivalent plastic shear strain (eps) obtained using the continuously updated elastic modulus, derived from real-time wave velocity inversion, are shown in Fig. 7.11a, b. For comparison, the eps distributions calculated using the conventional method, which assumes a constant elastic modulus, are displayed in Fig. 7.11c, d. The results in Fig. 7.11a, c correspond to cross-section A, ocated 2.5 m from auxiliary tunnel #2, as indicated in Fig. 7.11b, d.

The simulations indicate that significantly greater damage, both in magnitude and spatial extent, is predicted when the elastic modulus is updated dynamically. Field observations reveal that severe damage actually occurs in the vicinity of cross-section A, a condition that the conventional method fails to capture adequately (Fig. 7.11c, d). In contrast, the proposed method, by incorporating real-time updates of the elastic modulus, effectively overcomes this limitation and provides a more accurate representation of the damage distribution (Fig. 7.11a, b).

It is important to note that the excavation of tunnel B commenced only after auxiliary tunnel #2 had already been constructed. As a result, an excavation-damaged zone (EDZ) had already developed around tunnel #2 prior to the start of tunnel B excavation. This is evidenced by the monitored wave velocity field, particularly within the region indicated by the white dashed box in Fig. 7.11b. The formation of the EDZ induces stress concentration in areas farther from the tunnel, leading to greater compaction of the rock mass at distances of 4–10 m from tunnel #2. This phenomenon accounts for the localized increase in wave velocity observed in Fig. 7.11b.

In zones characterized by elevated wave velocity and correspondingly high elastic modulus, the capacity of the rock mass to store elastic energy is enhanced. This results in more pronounced elastoplastic deformation and a concomitant reduction in the degree of energy release. Such behavior explains the high–low–high–low zonal pattern of yield depth illustrated in Fig. 7.10a. Moreover, the trend of increased elastoplastic deformation in regions of high wave velocity and high elastic modulus is distinctly observable. For instance, a localized zone of elevated eps values appears near the tunnel sidewalls between the two dashed boxes in Fig. 7.11b. In contrast, Fig. 7.11d fails to capture this spatially varying pattern of elastoplastic deformation induced by the presence of tunnel #2. According to the conventional method, significant damage is predicted only in the immediate vicinity of auxiliary tunnel #2, with eps values falling below 0.003 elsewhere.

(2) Parameter Calculation

CASRock simulations rely on a comprehensive set of mechanical parameters. In this work, the elastic modulus is the sole parameter updated dynamically via the inverted P-wave velocity field; all other parameters are sourced from earlier research on the Jinping project (Jiang et al. 2010, 2016; Feng et al. 2020). The Jinping site has been the subject of extensive experimental and numerical study over many years, resulting in a robust understanding of its geological conditions and mechanical behavior. Hence, the parameters listed in Tables 7.2 and 7.3 are regarded as reliable inputs for the CASRock analysis.

(3) Sustainable development

The increasing demand for deep underground infrastructure has made cross-tunnel construction a common engineering scenario. The proposed method provides a means to analyze surrounding rock stability in such settings and to pinpoint areas of potential failure. Early identification enables timely reinforcement, thereby reducing accident risks and associated economic costs. Furthermore, the integration of disaster prevention measures helps mitigate environmental disruption. In this way, the method contributes to the sustainable development of society.

7.5 Conclusions

This work establishes a method for real-time stability assessment of surrounding rock in deep-buried cross-tunnels. Application in the JP tunnel supports the following conclusions:

(1) A novel approach is proposed for assessing the stability of the surrounding rock during tunnel excavation. This method integrates dynamic velocity field inversion via adjoint-state traveltime tomography with stability analysis conducted using CASRock software. By capturing real-time variations in the wave velocity field, the mechanical properties of the rock mass are continuously updated, thereby accounting for the progressive changes in elastic modulus associated with unloading during cross-tunnel excavation. Within the CASRock framework, the Mohr–Coulomb strength criterion and the cohesion weakening and friction strengthening model are incorporated. The extent and severity of rock damage are quantified using two key indicators: equivalent plastic shear strain (eps) and local energy release rate (LERR).

(2) The proposed method was applied to the excavation of the Jinping II hydropower station, yielding valuable insights into the zoning phenomena induced by the excavation process.

   a. The damage extent and severity progressively increase during the initial stages of excavation, with the most significant damage concentrated within the first 4.5 m of advance. The proximity to the auxiliary tunnel is strongly correlated with larger plastic zones and more intense damage.
   b. Spatial analysis of the yield depth distribution reveals that the largest yield depths are concentrated near the sidewalls, whereas those at the crown and invert are substantially smaller. The yield depth exhibits a distinctive pattern of variation with distance from the auxiliary tunnel, characterized by alternating high and low values. Peaks occur in the 0–2 m and 8–10 m intervals, a phenomenon potentially attributable to localized increases in wave velocity within these zones.

References

Chen X, Wang JG (2022) Stability analysis for compressed air energy storage cavern with initial excavation damage zone in an abandoned mining tunnel. J. Energy Storage 45:103725

Chen YX, Yong WX, Li CQ, Zhou J (2023) Predicting the thickness of an excavation damaged zone around the roadway using the DA-RF hybrid model. CMES-Comput. Model. Eng. Sci. 136:2507–2526

Cui Z, Sheng Q, Leng XL, Ma Y (2019) Investigation of the long-term strength of Jinping marble rocks with experimental and numerical approaches. Bull Eng Geol Environ 78:877–882

Czarny R, Malinowski M, Chamarczuk M, Cwiekata M, Olechowski S, Lsakow Z, Sierodzki P (2021) Dispersive seismic waves in a coal seam around the roadway in the presence of excavation damaged zone. Int J Rock Mech Min Sci 148:104937

Feng XT, Pan PZ, Zhou H (2006) Simulation of the rock microfracturing process under uniaxial compression using an elasto-plastic cellular automaton. Int J Rock Mech Min Sci 43:1091–1108

Feng XT, Xu H, Qiu SL, Li SJ, Yang CX, Guo HS, Cheng Y, Yao HG (2018) In situ observation of rock spalling in the deep tunnels of the China Jinping Underground Laboratory (2400 m depth). Rock Mech Rock Eng 51:1193–1213

Feng XT, Xu H, Yang CX, Zhang XW, Gao YH (2020) Influence of loading and unloading stress paths on the deformation and failure features of Jinping marble under true triaxial compression. Rock Mech Rock Eng 53:3287–3301

Feng XT, Wang ZF, Zhou YY, Yang CX, Pan PZ, Kong R (2021) Modelling three-dimensional stress-dependent failure of hard rocks. Acta Geotech 16:1647–1677

Feng XT, Pan PZ, Wang ZF, Zhang YL (2021) Development of cellular automata software for engineering rockmass fracturing processes. In: Challenges and innovations in geomechanics: proceedings of the 16th international conference of IACMAG-volume 1 16. Springer, Berlin/Heidelberg, Germany, pp 62–74

Guo HS, Feng XT, Li SJ, Yang CX, Yao ZB (2017) Evaluation of the integrity of deep rock masses using results of digital borehole televiewers. Rock Mech Rock Eng 50:1371–1382

Hadjigeorgiou J, Esmaieli K, Grenon M (2009) Stability analysis of vertical excavations in hard rock by integrating a fracture system into a PFC model. Tunn Undergr Space Technol 24:296–308

Hajiabdolmajid V, Kaiser PK, Martin CD (2002) Modelling brittle failure of rock. Int J Rock Mech Min Sci 39:731–741

He Y, Liu Y, Hazarika H, Yuan R (2019) Stability analysis of seismic slopes with tensile strength cut-off. Comput Geotech 112:245–256

Hu JT, Qian JL, Cao JX, Wang XJ, Wang HZ, Leung SY (2021) Ray-illumination compensation for adjoint-state first-arrival traveltime tomography. Geophysics 86:U109–U119

Jiang Q, Feng XT, Xiang TB, Su GS (2010) Rockburst characteristics and numerical simulation based on a new energy index: a case study of a tunnel at 2500 m depth. Bull Eng Geol Environ 69:381–388

Jiang Q, Zhong S, Cui J, Feng XT, Song LB (2016) Statistical characterization of the mechanical parameters of intact rock under triaxial compression: An experimental proof of the Jinping marble. Rock Mech Rock Eng 49:4631–4646

Leparoux D, Côte P, Gélis C (2012) EDZ characterization with surface wave analysis: An experimental and numerical study for defining feasibility in the context of the Tournemire platform (France). Near Surf Geophys 10:401–411

Lin Y, Ma J, Lai Z, Huang L, Lei M (2023) A FDEM approach to study mechanical and fracturing responses of geo-materials with high inclusion contents using a novel reconstruction strategy. Eng Fract Mech 282:109171

Liu L, Li SJ, Xiao YX, Chen SJ, Wang ZF, Feng GL, Wang Y (2023) An inverted heterogeneous velocity model for microseismic source location in deep buried tunnels. Rock Mech Rock Eng 56:4855–4880

Liu L, Wang Y (2021) Perturbational inversion of fundamental mode Rayleigh wave dispersion curve. In: Proceedings of the SEG international exposition and annual meeting, Denver, Colorado, USA, September 26–October 1, p D011S075R007

Ma J, Chen J, Chen W, Huang L (2022) A coupled thermal-elastic-plastic-damage model for concrete subjected to dynamic loading. Int J Plast 153:103279

Ma J, Zhao J, Lin Y, Liang J, Chen J, Chen W, Huang L (2023) Study on tamped spherical detonation-induced dynamic responses of rock and PMMA through mini-chemical explosion tests and a four-dimensional lattice spring model. Rock Mech Rock Eng 56:7357–7375

Müller W (1991) Numerical simulation of rock bursts. Min Sci Technol 12:27–42

Ozaki Y (2023) First-arrival traveltime tomography for monitoring the excavation damaged zone in the Horonobe Underground Research Laboratory. Rock Mech Bull 2:100057

Pan PZ, Feng XT, Hudson JA (2006) Numerical simulations of Class I and Class II uniaxial compression curves using an elasto-plastic cellular automaton and a linear combination of stress and strain as the control method. Int J Rock Mech Min Sci 43:1109–1117

Park CB, Miller RD, Xia J (1998) Imaging dispersion curves of surface waves on multi-channel record. In SEG technical program expanded abstracts 1998. Society of Exploration Geophysicists: Tulsa, OK, USA, pp 1377–1380

Sakurai S (2017) Back analysis in rock engineering. CRC Press, Boca Raton FL USA

Song Z, Liu H, Kong X (2021) Development of excavation damaged zones around a rectangular roadway under mining-induced pressure. Tunn Undergr Space Technol 118:104163

Taillandier C, Noble M, Chauris H Calandra H (2009) First-arrival traveltime tomography based on the adjoint-state method. Geophysics 74 (WCB1–WCB10)

Verruijt A, Booker JR (1998) Surface settlements due to deformation of a tunnel in an elastic half plane. Geotechnique 48:709–713

Walton G (2019) Initial guidelines for the selection of input parameters for cohesion-weakening-friction-strengthening (CWFS) analysis of excavations in brittle rock. Tunn Undergr Space Technol 84:189–200

Wang JA, Park HD (2001) Comprehensive prediction of rockburst based on analysis of strain energy in rocks. Tunn Undergr Space Technol 16:49–57

Xiao Y, Chen S, Liu L et al (2024) Determining S-wave arrival times for microseismic monitoring in tunnels based on instantaneous phase difference of sensors. Rock Mech Rock Eng 57(3):1781–1797

Xiao Y, Chen S, Feng G et al (2025) Picking P-wave arrival times from microseismic signals in tunnels with low signal-to-noise ratios without noise filtering. Soil Dyn Earthq Eng 192:109288

Zhu WC, Wei J, Zhao J, Niu LL (2014) 2D numerical simulation on excavation damaged zone induced by dynamic stress redistribution. Tunn Undergr Space Technol 43:315–326

Zhu Z, Ranjith PG, Tian H, Jiang G, Dou B, Mei G (2021) Relationships between P-wave velocity and mechanical properties of granite after exposure to different cyclic heating and water cooling treatments. Renew Energy 168:375–392

Chapter 8
Stratified Double-Difference Dynamic Tomography of a 3D Velocity Field for a Deep Buried Large Underground Cavern

8.1 Prologue

Among methods to assess tunnel stability (Li et al. 2013), velocity fields provide insights into geological structures, excavation-induced damage, and stress distribution in rock masses. Dynamic velocity variations can reflect stress evolution and identify high-stress regions (Xiao et al. 2023). Thus, dynamic inversion of the velocity field is urgently needed in large underground cavern projects.

Inversion methods can generally be divided into two categories (active and passive methods) based on the input data. If the triggering time and coordinates of the source are known, it is considered active. A typical example of an active seismic method is prestack velocity analysis, utilized in tunnel seismic prediction (TSP) for constructing velocity models (Lin and Li 2014). This method continuously refines the velocity model by analyzing traveltime changes with offset, thus affecting the efficiency of normal moveout (NMO) correction. By achieving flat events post NMO, a velocity model that precisely mirrors the subsurface structure can be acquired (Shih and Chen 1996). Despite TSP showing acceptable performance in linear engineering scenarios such as tunnels, it is still unsuitable for deeply buried caverns (Lu et al. 2024). This approach lacks adequate resolution to meet engineering requirements, as it only fits relatively straightforward geological settings (Xue et al. 2019). Hence, inversion methods that are able to produce high-resolution velocity fields are essential.

With the progress in computational hardware, full waveform inversion (FWI) has become a method with high precision and high resolution (Lin et al. 2023). FWI uses entire wavefield information to invert medium parameters, yet its inversion process is a highly ill-posed nonlinear issue. Tackling the problems of nonunique solutions and convergence is still a major challenge in FWI applications, which has led many scholars to propose creative solutions (Bunks et al. 1995; Pratt 1990). FWI has been progressively introduced into tunnel settings. Researchers have devised optimization strategies for specific modules of the FWI workflow to boost computational

© Tongji University Press Co., Ltd 2026

Y. Xiao et al., *Near-Field Microseismic Data Processing in Deep Tunnels*,
https://doi.org/10.1007/978-981-92-3208-6_8

efficiency, reduce memory needs, enhance inversion accuracy, and ensure applicability specific to tunnels (Liu et al. 2019). For instance, Musayev et al. (2016) and Yu et al. (2022) employed frequency-domain FWI in tunnels, whereas Jiang et al. (2024) integrated deep learning with FWI to create a new inversion framework. Nevertheless, FWI demands data produced by active seismic sources with accurately known triggering times, precise estimation of source wavelets, and considerable computational resources. In tunnels, it is unfeasible to provide a great number of active sources over long periods, making FWI unsuitable for time-lapse monitoring of velocity fields in deeply buried engineering.

Traveltime tomography, which similarly depends on active seismic sources, works by iteratively refining slowness models to attain the best possible match between simulated and observed traveltimes (Dong et al. 2023). The primary calculations consist of updating the slowness model through iterative inversion and precisely determining ray paths using forward modeling (Chen et al. 2023). Multiple ray-tracing algorithms, such as the bending method (Koketsu and Sekine 1998), shooting method (Kasdorf et al. 2021), and shortest-path method (Bai et al. 2007), have been utilized in various geological settings. Recent progress in adjoint-state formulations has greatly improved convergence and efficiency, especially when dealing with nonlinear objective functions in intricate velocity structures (Wu et al. 2025; Liu et al. 2023). Although traveltime tomography also necessitates active seismic sources, its merit is in fulfilling engineering demands while streamlining the computational process. In contrast to FWI, it maintains the unique advantage of computational efficiency while meeting practical engineering needs.

Using passive seismic sources such as MS events, blasting sources, construction noise, and earthquake data for traveltime tomography is more appropriate for acquiring high-resolution time-lapse velocity models of deeply buried caverns. Double-difference (DD) traveltime tomography with earthquake data has been extensively employed (Zhang and Thurber 2003). This method tackles the issue of unknown source parameters by using the double difference between observed and calculated traveltimes of event pairs. In small-scale underground engineering, MS events can act as an ideal data source for inversion. Research has indicated that many MS events are produced in rocks during or after cavern excavation (Lai et al. 2022), which makes it feasible to use MS events for DD tomography.

Despite the DD method having made remarkable achievements in large-scale geophysical applications (Zhang et al. 2004), its application in small-scale underground engineering faces issues. This restriction arises from the distinctions in source-station arrangements (Liu et al. 2020). In large-scale setups, seismic sources are usually situated beneath the receiver array along the depth axis. In contrast, in deeply buried projects, MS sensors are frequently placed among the sources. The traditional pseudobending algorithm in the DD method for forward traveltime computation is not appropriate for such source-receiver layouts, causing considerable deviations in calculated traveltimes during inversion and eventually making the inversion ineffective.

This chapter introduces a stratified double-difference (SDD) tomography method to address these issues, which successfully reduces divergence problems triggered by uneven source-sensor distributions. The method takes MS events as input for deriving three-dimensional velocity fields and their changes within large underground caverns. Using the Jinping large underground laboratory project as an example, two checkerboard models are created to assess the resolution of the suggested method, one including the excavated Jinping laboratory structures and the other not. Dynamic inversion of the velocity field during excavation is subsequently carried out for the Jinping laboratory. Last, the effect of stratified versus nonstratified inversion on velocity outcomes is examined, together with the greatest acceptable input data error for the suggested method.

8.2 Methodology

8.2.1 *Stratified Double-Difference Tomography Method (SDD)*

Figure 8.1 demonstrates the overall workflow of the velocity field inversion method. The process consists of four key parts: data preprocessing, data quality control, stratification and clustering, and velocity field inversion.

First, the absolute P-wave arrival time of each MS signal is extracted. An initial estimate of the rock mass's P-wave velocity is obtained based on existing geological data or artificial sources. Using this estimated velocity, the locations and origin times of the MS events are determined. Signals with excessive discrepancies in epicentral distance or travel time relative to predefined thresholds are identified and removed as outliers.

The inversion strategy then involves dividing the model into upper and lower submodels. Within each submodel, the MS events are clustered using the density-based spatial clustering of applications with noise (DBSCAN) method to select those suitable for inversion. These data selection steps effectively reduce data redundancy and mitigate the adverse influence of outliers on the inversion results.

Subsequently, the velocity model is updated using data from paired MS events. The forward-calculated traveltime differences obtained via a pseudobending ray-tracing algorithm are compared with the observed traveltime differences. The inversion process iterates until an optimal match between the calculated and observed values is achieved, at which point the final velocity field is output. Detailed descriptions of each step are provided in the following sections.

A. **Data Preprocessing**

Reliable velocity inversion relies on the precise extraction of P-wave absolute arrival times. Nevertheless, the vast amount of MS data renders manual picking extremely time-consuming and labor-intensive. Consequently, an automatic method for picking P-wave arrival times is needed. This research employs a technique based on the instantaneous phase difference among signals to ascertain P-wave arrival times (Xiao

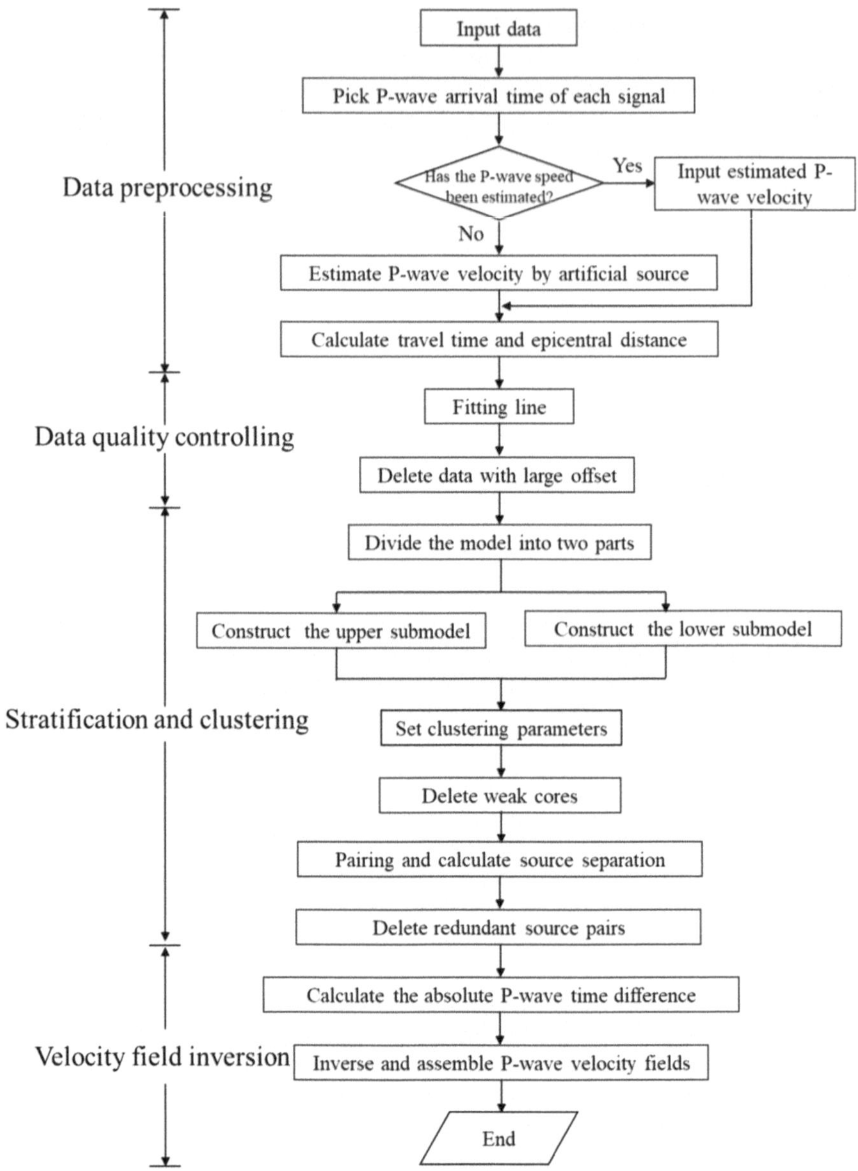

Fig. 8.1 Workflow of SDD

et al. 2024 and Xiao et al. 2025). This method efficiently diminishes the impact of noise on picking precision, thus guaranteeing the high-accuracy extraction of P-wave arrival times.

If an estimated P-wave velocity is available from project documentation, it can be directly used in subsequent analyses. In cases where such data are insufficient, a single active source may be deployed to obtain an approximate P-wave velocity. This is achieved by calculating the average velocity based on the propagation distance and corresponding traveltime. Using the observed P-wave arrival times, the coordinates of the sensor array, and the estimated velocity, the spatial coordinates and origin time of each MS event can then be determined.

B. **Data Quality Control**

Given that manual and automated P-wave picks in MS monitoring are unavoidably affected by minor errors (Jia et al. 2021), robust quality control, centered on the elimination of outlier arrivals, is necessary to guarantee the reliability of subsequent velocity inversion.

The MS wave traveltime t can be derived through the relationships below:

$$t = T - \tau \tag{8.1}$$

Here, T denotes the observed P-wave arrival time, and τ stands for the triggering time. For each sensor channel, the epicentral distance d is computed. The specific procedures for outliers are as follows:

(1) Conduct a least-squares linear regression analysis on the t-d dataset to obtain the best-fit line $y = sl{\cdot}x + o$, with sl and o being the slope and intercept parameters, respectively.
(2) Set an intercept tolerance threshold o_r build upper and lower boundary lines.

$$y = sl \cdot x + o \pm o_\mathrm{r} \tag{8.2}$$

(3) For every measured d value, calculate the corresponding y_1 and y_2 via Eq. 8.2. Keep the observations that meet $y_2 \leq y \leq y_1$ and discard those outside this bounded range.

C. **Stratification and Clustering**

Model stratification and MS event clustering constitute essential steps in the SDD tomography workflow. Omitting the stratification step can substantially reduce the accuracy of the velocity inversion, as further discussed in Sect. 8.6. Event clustering serves to systematically eliminate anomalous signals, thereby preserving the stability of the inversion process (Zhang and Thurber 2006).

The model is divided into upper and lower submodels at a depth corresponding to the central elevation of the underground excavation. The dividing boundary is extended to ensure that all sensors are encompassed within the two submodels, resulting in certain depth intervals being represented in both submodels simultaneously. It is recommended to adopt the mean depth of the sensors as the stratification plane, set at $Z = 80$ m in this study, which helps maintain a relatively balanced overlap of sensors and sources between the upper and lower submodels.

This strategy effectively mitigates the adverse effects of the uneven spatial distribution of sensors and sources on velocity field inversion, which is particularly beneficial in deep underground engineering contexts.

The DBSCAN method exhibits strong performance in detecting outliers among irregularly distributed discrete sources (Deng 2020), making it well suited to the requirements of the SDD method. The algorithm proceeds by sequentially treating each MS source as a core object. For each core, a spatial searching radius α is defined (as illustrated in Fig. 8.2), and a density threshold β is used to classify sources based on the number of neighboring points within that radius. Sources with more than β neighbors within the radius are designated strong cores, while those with fewer neighbors are considered weak cores. This classification effectively filters out distant sources, retaining only strong cores for subsequent tomographic inversion.

The choice of parameters α and β is influenced by two factors: the spatial distribution features of the MS events and the planned extent of the inversion. In the checkerboard tests carried out in this research, relatively loose standards were used because the MS events were broadly spread and a global inversion was needed for resolution analysis (Sect. 8.4). On the other hand, stricter standards were employed in the inversion tests based on the Jinping field conditions (Sect. 8.5), where MS events were more clustered near the caverns and the inversion concentrated on local velocity information. It is worth mentioning that excessively rigorous standards might lead to an inadequate number of eligible data points, thus restricting inversion coverage or even making the inversion unworkable.

Match the N strong cores that have been preserved, which theoretically produces C_N^2 pairs. To enhance computational efficiency and reduce data redundancy, a

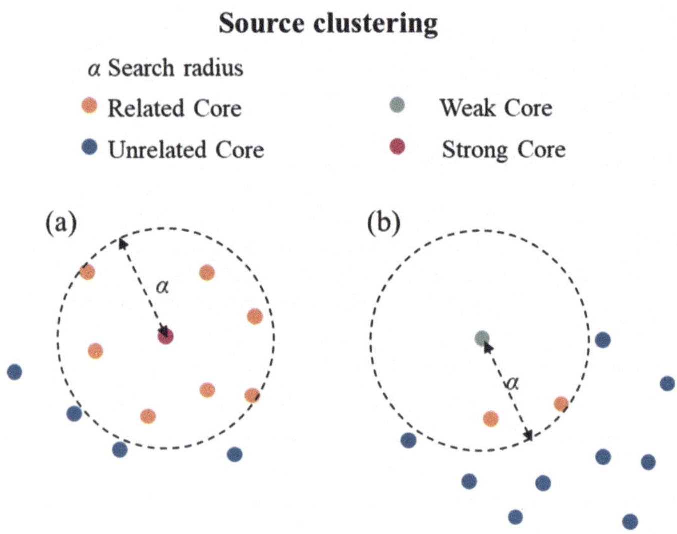

Fig. 8.2 Diagram of the DBSCAN algorithm. Identification of **a** strong core sources and **b** weak core sources

distance threshold R_{dmax} is established to rule out source pairs surpassing the maximum permissible separation. Subsequently, sensor selection criteria (Num_{smax} and Num_{smin}) are utilized for every source pair. Pairs with less than Num_{smin} common sensors are eliminated, and those surpassing Num_{smax} experience sensor prioritization according to source-sensor distance to remove superfluous common sensors. Ultimately, the absolute P-wave arrival time difference is computed for all eligible pairs.

D. **Velocity Field Inversion**

The inversion process is initialized with four sets of input data: the coordinates of the MS sensors, the absolute P-wave arrival time differences for each event pair, the estimated source locations, and the corresponding triggering times. An initial homogeneous velocity field is constructed based on the estimated P-wave velocity. The traveltimes between each source and sensor are then computed using the pseudobending ray-tracing algorithm (Koketsu and Sekine 1998). These calculated traveltimes are iteratively compared with the observed values. The inversion solves for velocity updates by minimizing the L2 norm of the traveltime differences, employing the least-squares QR-factorization (LSQR) algorithm (Yao et al. 1999).

Following each update of the velocity field, a relocation of the MS events is performed using the double-difference relocation method (Waldhauser and Ellsworth 2000). This approach compares the differential traveltimes of event pairs recorded at common stations, effectively canceling common errors associated with the source region and the ray paths to that station. As a result, the influence of velocity model inaccuracies is significantly reduced, enabling high-precision source locations. The updated source coordinates are then fed back into the ray-tracing module for subsequent iterations. The iterative process continues until an optimal match between the calculated and observed traveltime differences is achieved, at which point the final velocity field is output.

8.2.2 Double-Difference Traveltime Tomography

This section elaborates on the velocity inversion algorithm. The absolute P-wave arrival time T for MS event i, recorded by sensor k, can be formulated as:

$$T_k^i = \tau^i + \int_i^k u\mathrm{d}s \tag{8.3}$$

τ^i represents the origin time of MS event i. u and s denote the slowness field and the propagation path, respectively. The location and origin time of the MS event, the propagation path, and the velocity of the surrounding rock medium are all undetermined. Furthermore, the relationship between the arrival time and the event location is nonlinear. Consequently, a Taylor expansion is employed to linearize Eq. 8.3, and simultaneously, the connection among the absolute arrival time, the predicted arrival time ${r_k}^i$, and the velocity structure parameters under a three-dimensional (3D) velocity model is established.

$$r_k^i = \sum_{l=1}^{3} \frac{\partial T_k^i}{\partial x_l^i} \Delta x_l^i + \Delta\tau^i + \sum_{m=1}^{M_{ik}} \sum_{n=1}^{N} \omega_{mn} \Delta u_n \Delta s_m \tag{8.4}$$

The ray path from source i to sensor k is divided into M_{ik} segments, each having a length of Δs_m, and an interpolation weight ω_{mn} is assigned to each adjacent velocity grid. According to the pseudobending algorithm (Junho and Clifford 1987), the curved ray is split into multiple short straight segments. The number of segments in a ray path is adjustable. The pseudobending ray-tracing algorithm continues its iteration for ray searching until it locates the best path. This kind of ray tracing will be utilized in the new velocity field after every iterative update. Since the velocity field is discretized on a 3D grid, the travel time is acquired via numerical summation along the ray-path segments in line with the trapezoidal rule. For ray segments whose end points are not on the velocity grid, the velocities at the end points are obtained through the trilinear interpolation method based on that of the nearby grid points. The value of the weight ω_{mn} is decided by the distance from the midpoint of the ray segment to the neighboring velocity grid points according to this interpolation method. For more detailed equations and calculation methods, one can refer to Junho and Clifford (1987) and Zhang and Thurber 2003, 2006.

Another MS event j also meets Eq. 8.4. Subtract them to obtain:

$$r_k^i - r_k^j = \sum_{l=1}^{3} \frac{\partial T_k^i}{\partial x_l^i} \Delta x_l^i + \Delta\tau^i + \sum_{m=1}^{M_{ik}} \sum_{n=1}^{N} \omega_{mn} \Delta u_n \Delta s_m - \sum_{l=1}^{3} \frac{\partial T_k^j}{\partial x_l^j} \Delta x_l^j + \Delta\tau^j - \sum_{m=1}^{M_{jk}} \sum_{n=1}^{N} \omega_{mn} \Delta u_n \Delta s_m \tag{8.5}$$

Eq. 8.5 can be simplified as

$$r_k^i - r_k^j = \left(T_k^i - T_k^j\right)^{\text{obs}} - \left(T_k^i - T_k^j\right)^{\text{cal}} \tag{8.6}$$

where $(T_k{}^i - T_k{}^j)^{\text{obs}}$ and $(T_k{}^i - T_k{}^j)^{\text{cal}}$ represent the observing time difference and the calculating time difference, respectively.

If every MS event has N_{p} arrivals (a collection of p=1, 2, …, P events), Eq. 8.4 can be converted into the following format.

$$\boldsymbol{A}\Delta\boldsymbol{X} + \boldsymbol{C}\Delta\boldsymbol{M} + \boldsymbol{S} = \Delta\boldsymbol{T} \tag{8.7}$$

where $\boldsymbol{A}$ represents the partial derivative matrix related to the hypocenter and origin time, $\boldsymbol{\Delta X}$ denotes the perturbation vector for earthquake location and origin time, $\boldsymbol{C}$ refers to the model derivative matrix associated with the slowness model, $\boldsymbol{\Delta M}$ is the vector of slowness perturbations, $\boldsymbol{\Delta T}$ indicates the misfit of arrival time, and $\mathbf{S}$ stands for the station correction vector. After every update of the velocity field, the sources are relocated using the double-difference location method (Waldhauser and Ellsworth 2000). $\boldsymbol{A}$ is refreshed after ray tracing according to the latest velocity field and source locations, and $\boldsymbol{C}$ is calculated by multiplying the interpolation weights by the lengths of the ray segments. For more detailed information on the equation, refer to Zhang and Thurber 2003, 2006.

The matrices are presented below:

$$\boldsymbol{A} = \begin{bmatrix} A_1 & 0 & \cdots & 0 \\ 0 & A_2 & \cdots & 0 \\ \vdots & \vdots & \ddots & \vdots \\ 0 & 0 & \cdots & A_P \end{bmatrix}; N_t \times M_t \tag{8.8}$$

$$\Delta \boldsymbol{T} = \begin{bmatrix} \Delta T_1 \\ \Delta T_2 \\ \vdots \\ \Delta T_P \end{bmatrix}; N_t \times 1 \tag{8.9}$$

$$\boldsymbol{C} = \begin{bmatrix} C_1 & 0 & \cdots & 0 \\ 0 & C_2 & \cdots & 0 \\ \vdots & \vdots & \ddots & \vdots \\ 0 & 0 & \cdots & C_P \end{bmatrix}; N_t \times L \tag{8.10}$$

$$\Delta \boldsymbol{M} = \begin{bmatrix} \Delta m_1 \\ \Delta m_2 \\ \vdots \\ \Delta m_L \end{bmatrix}; L \times 1 \tag{8.11}$$

$$\boldsymbol{S} = \begin{bmatrix} s_1 \\ s_2 \\ \vdots \\ s_P \end{bmatrix}; N_t \times 1 \tag{8.12}$$

where $N_t = \sum_{p=1}^{P} N_p$ is the total number of arrival time data and $M_t = 4 \times P$ is the total number of unknown origin parameters.

Adding a differential operator $\boldsymbol{Q}_{\mathbf{DD}}$ to Eq. 8.7 yields the matrix operation form of Eq. 8.5:

$$\boldsymbol{Q}_{\mathbf{DD}}\boldsymbol{A}\,\Delta\boldsymbol{X} + \boldsymbol{Q}_{\mathbf{DD}}\boldsymbol{C}\,\Delta\boldsymbol{M} = \boldsymbol{Q}_{\mathbf{DD}}\,\Delta\boldsymbol{T} \tag{8.13}$$

When $\boldsymbol{Q}_{\mathbf{DD}}\boldsymbol{S}$ equals 0 (Zhang and Thurber 2006), the double difference location algorithm fails to account for the abnormal path deviations between events. Thus, it can only choose events that are in closer proximity (Wolfe 2002). Nevertheless, this tomography algorithm specifically addresses this matter. Moreover, all potential combinations of arrival times for shared sensors are considered, ensuring that the matrix $\boldsymbol{Q}_{\mathbf{DD}}\boldsymbol{A}$ achieves full column rank. If the parameters of the slowness model not sampled via the ray path are excluded during inversion, then $\boldsymbol{Q}_{\mathbf{DD}}\boldsymbol{C}$ will also attain full column rank. The formula for calculating the velocity field using absolute arrival time data and differential data is derived from Eq. 8.13 as follows:

$$\boldsymbol{Y} = \begin{pmatrix} \boldsymbol{Q}_{\mathbf{DD}}\boldsymbol{E} \\ \omega \boldsymbol{E} \end{pmatrix}^{-1} \begin{pmatrix} \boldsymbol{Q}_{\mathbf{DD}} \\ \omega \boldsymbol{I} \end{pmatrix} \Delta \boldsymbol{T} \tag{8.14}$$

$$\boldsymbol{E} = \begin{bmatrix} \boldsymbol{A} & \boldsymbol{C} \end{bmatrix} \tag{8.15}$$

$$\boldsymbol{Y} = \begin{bmatrix} \Delta \boldsymbol{X} \\ \Delta \boldsymbol{M} \end{bmatrix} \tag{8.16}$$

where ω denotes the relative weight of the absolute arrival time and relative arrival time, and **I** represents the identity matrix.

8.3 MS Monitoring Setup for the Jinping Underground Laboratory

This study evaluates the performance of the proposed velocity tomography method using a realistic MS monitoring configuration deployed at the Jinping underground laboratory (Fig. 8.3). The background information can be found in Sect. 6.4.

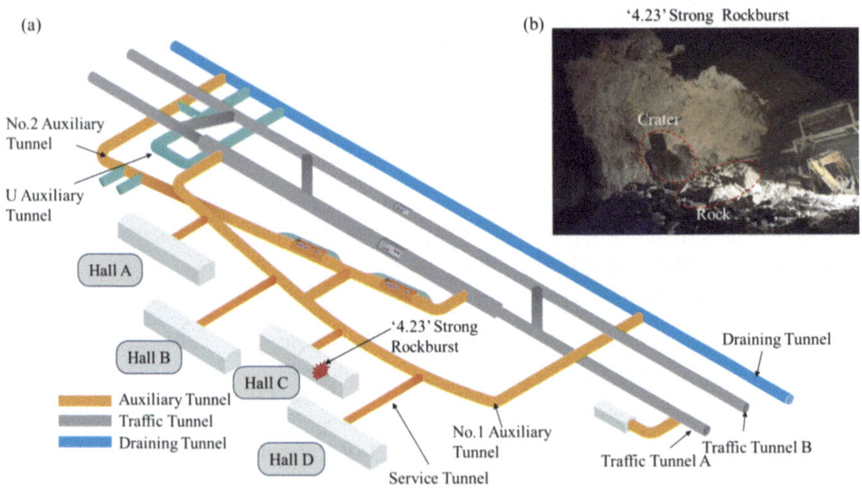

Fig. 8.3 Background of the JP underground laboratory. **a** Underground laboratory model. **b** Photo of the "4.23" strong rockburst

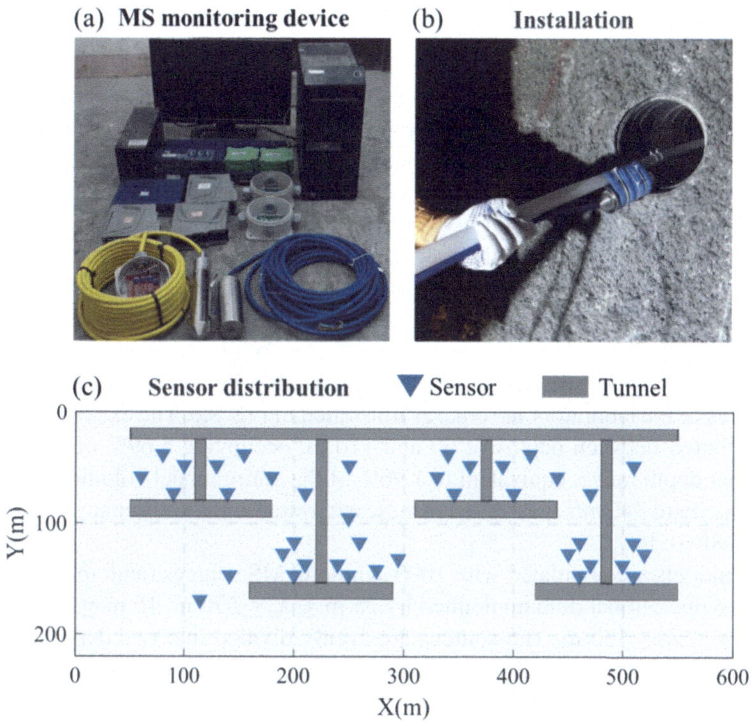

Fig. 8.4 MS sensor configuration **a** photograph of the monitoring instrument; **b** installation of sensors in the field; **c** sensor arrangement in the Jinping underground laboratory

To ensure high-quality signal acquisition, boreholes ranging from 3 to 25 m in depth were drilled into the rock walls. Sensors were installed at the deepest point of each borehole, and coupling material was used to guarantee faithful recording of vibration signals from the surrounding rock. A total of 40 sensors, each with a sampling frequency of 4000 Hz, were deployed around the four laboratory caverns and the service tunnels, as illustrated in Fig. 8.4. This spatial configuration establishes an effective sensor array system that enhances both monitoring accuracy and coverage across the entire area of interest.

8.4 Resolution Test

This section examines the performance of the SDD method through two checkerboard tests, concentrating on its resolution in two situations. One checkerboard model includes excavated laboratory structures, and the other model does not. Both are founded on the Jinping MS monitoring setup.

8.4.1 Setup for the Test

This section presents the construction of two 3D checkerboard models designed to evaluate the resolution of the proposed velocity tomography method. The first model (Model-1) spans 600 m in the X-direction, 220 m in the Y-direction, and 160 m in the Z-direction and incorporates a checkerboard velocity pattern composed of 50 m cubic blocks (Fig. 8.5). Alternating high-velocity (6000 m/s) and low-velocity (5000 m/s) zones are assigned throughout the model, with this pattern maintained uniformly at all depths. Notably, Model-1 does not include laboratory cavern structures.

A second checkerboard model (Model-2) is then constructed to represent the Jinping underground laboratory more realistically. While preserving the identical checkerboard velocity pattern used in Model-1, this model integrates the actual geometries of the laboratory caverns, as illustrated in Fig. 8.6. The excavated regions are distributed between depths of 60 and 110 m, occupying 5.69% of the volume within that depth range, equivalent to 1.96% of the entire model volume. A reduced wave velocity of 340 m/s is assigned to these excavated areas to simulate the presence of air-filled voids.

Both models are populated with 1000 synthetic MS sources randomly generated within the operational domain defined by 25 m $\leq X \leq$ 575 m, 15 m $\leq Y \leq$ 210 m, and 10 m $\leq Z \leq$ 150 m. The sources are evenly divided into two depth intervals: 500 events between 10 and 80 m and 500 events between 80 and 150 m. In Model-2, sources are deliberately excluded from the excavated regions to ensure that they are located exclusively within the rock mass. The same sensor configuration as in the Jinping field setup, comprising 40 sensors, is used in both models to record synthetic P-wave arrival times.

Under idealized conditions, the inversion is performed using the true source locations and precisely calculated P-wave arrival times as input, deliberately avoiding any estimation errors. Forward simulation of wave propagation in the checkerboard velocity models is carried out by solving the eikonal equation using the Multi-stencils Fast Marching Method (MSFM) (Hassouna and Farag 2007). As an advanced variant

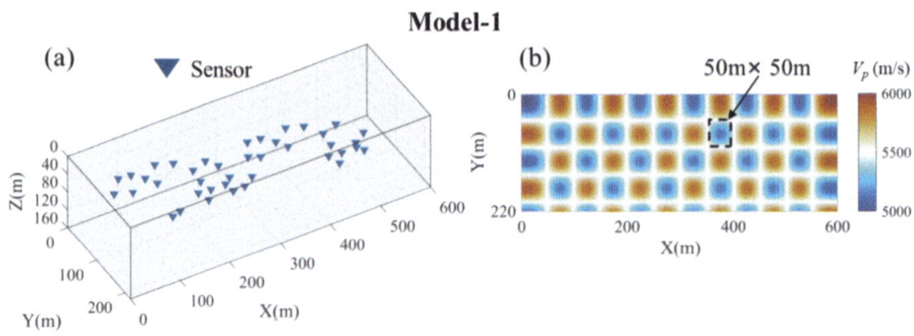

Fig. 8.5 Checkerboard model without embedded structures. **a** 3D model geometry; **b** true P-wave velocity at various depth levels

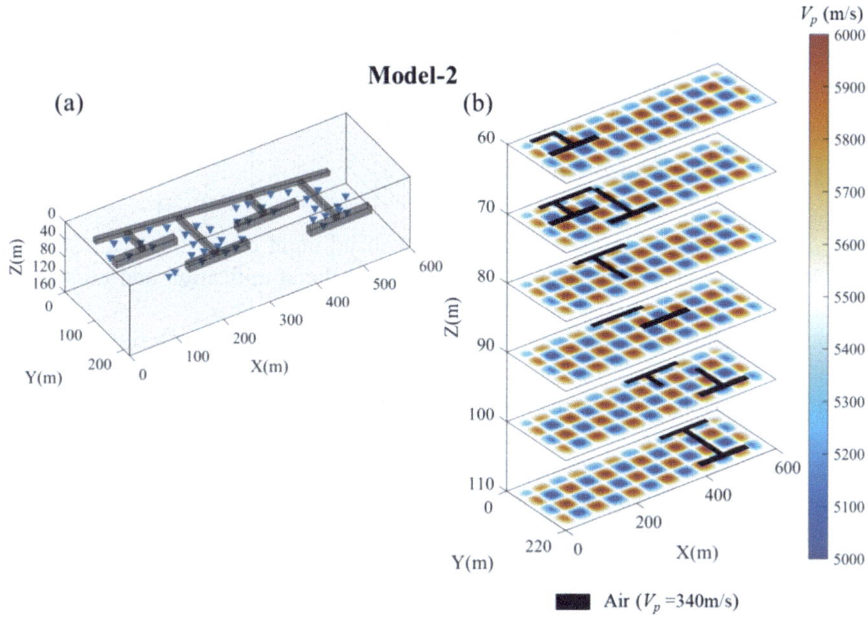

Fig 8.6 Checkerboard model of the Jinping underground laboratory. **a** 3D model geometry; **b** true P-wave velocity distribution at different depths

of the traditional fast marching method (FMM) (Sethian and Popovici 1999), the MSFM offers superior adaptability to complex models while maintaining high accuracy. For synthetic data generation, the MSFM is employed due to its precision, although its computational speed is relatively slower (Qu et al. 2014). Within the inversion itself, ray tracing is performed using the pseudobending algorithm (Koketsu and Sekine 1998), which provides the rapid computational speed necessary to ensure inversion efficiency. Stratification and clustering analyses are conducted directly on the synthetic datasets without additional data quality control steps.

8.4.2 The Evaluation Criterion for Inversion

The inversion performance can be quantitatively assessed through normalized cross-correlation (*CC*) analysis between the true velocity **Tr** and the inverted velocity **In**. The formula is presented as follows (Haralick et al. 1992).

$$CC = \frac{\sum_x \sum_y \left(\mathrm{Tr}_{xy} - \overline{\mathrm{Tr}}\right)\left(\mathrm{In}_{xy} - \overline{\mathrm{In}}\right)}{\sqrt{\sum_x \sum_y \left(\mathrm{Tr}_{xy} - \overline{\mathrm{Tr}}\right)^2 \sum_x \sum_y \left(\mathrm{In}_{xy} - \overline{\mathrm{In}}\right)^2}} \tag{8.17}$$

CC can evaluate the similarity between two matrices within − 1 and 1. The *CC* value increases with increasing similarity.

8.4.3 Velocity Model Inversion

Data quality control is unnecessary under idealized input conditions. This section uses Model-1 to provide a detailed illustration of the stratification and clustering procedures.

The model consists of upper and lower submodels, each covering the entire sensor array. The vertical ranges for the upper and lower submodels are 0–120 m and 40–160 m, respectively, with an overlap region between 40 and 80 m depth shared by both submodels (Fig. 8.7). A separation plane at ($Z = 80$ m) (shown as the red plane in Fig. 8.7) is utilized to distinguish the synthetic sources.

Cluster analysis is then performed on each submodel using the parameters specified in Table 8.1. The clustering results identify 484 strong core sources in the upper submodel and 473 in the lower submodel, yielding a total of 957 strong core sources. Correspondingly, 16 weak core sources are identified in the upper submodel and 27 in the lower submodel. Figure 8.8 illustrates the depth distribution of both source types. It should be noted that synthetic sources were not generated within the depth ranges of 0–10 m and 150–160 m.

The initial velocity model for inversion is defined as a homogeneous field with a constant velocity of 5,500 m/s (Fig. 8.9a), corresponding to the average of the high-velocity (6000 m/s) and low-velocity (5000 m/s) checkerboard values. Owing to the absence of MS sources in the 0–10 m and 150–160 m depth intervals, velocity updates within these ranges are excluded from the inversion. The inversion results are presented in Fig. 8.9b. Considering the spatial bounds of source generation, the effective inversion domain is confined to the region defined by 25 m $\leq X \leq$ 575 m and 15 m $\leq Y \leq$ 210 m.

8.4.4 Inversion Results and Analysis

(1) Checkerboard Without Laboratory Structures

Figure 8.9b, c display the inversion results obtained for Model-1. The reconstructed velocity structure closely reproduces the checkerboard pattern, with an average *CC* of 0.84. A notable reduction in *CC* is observed within the depth ranges of 10–40 m and 120–150 m, whereas the *CC* remains relatively stable between 40 m and 120 m. Interestingly, within the shallower (10–40 m) and deeper (120–150 m) intervals, the inversion performs better in recovering low-velocity anomalies than in recovering high-velocity anomalies. These results demonstrate that the SDD method achieves

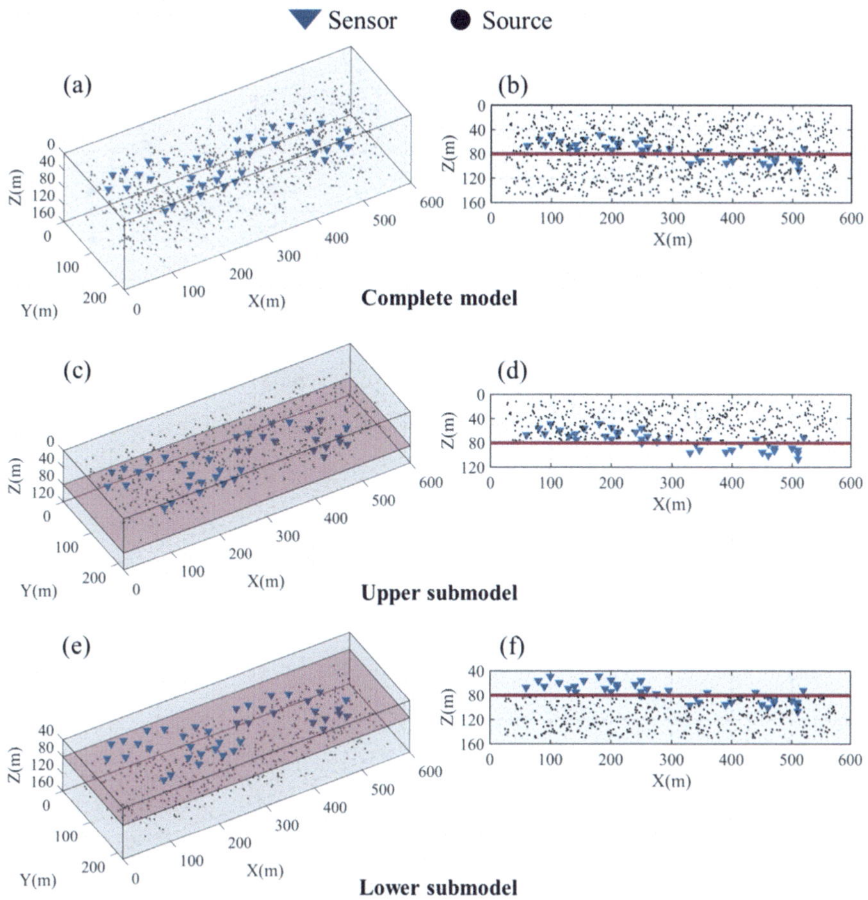

Fig. 8.7 Stratified model geometry. Perspective views: **a** complete model, **c** upper submodel, **e** lower submodel. Lateral views: **b** complete model, **d** upper submodel, **f** lower submodel

Table 8.1 Parameters for the DBSCAN method

Parameter	α (m)	β	R_{dmax} (m)	Num_{smin}	Num_{smax}
Value	45	8	150	8	20

excellent performance and reliability in velocity field inversion under ideal data conditions.

(2) Checkerboard Containing Laboratory Structures

Figure 8.10 displays the inversion results for the realistic velocity field shown in Fig. 8.6. The initial velocity model utilized in this inversion process (Fig. 8.10a) matches the one applied to Model-1, with no inclusion of the laboratory cavern structures. A comparison with the Model-1 outcomes reveals that the existence of

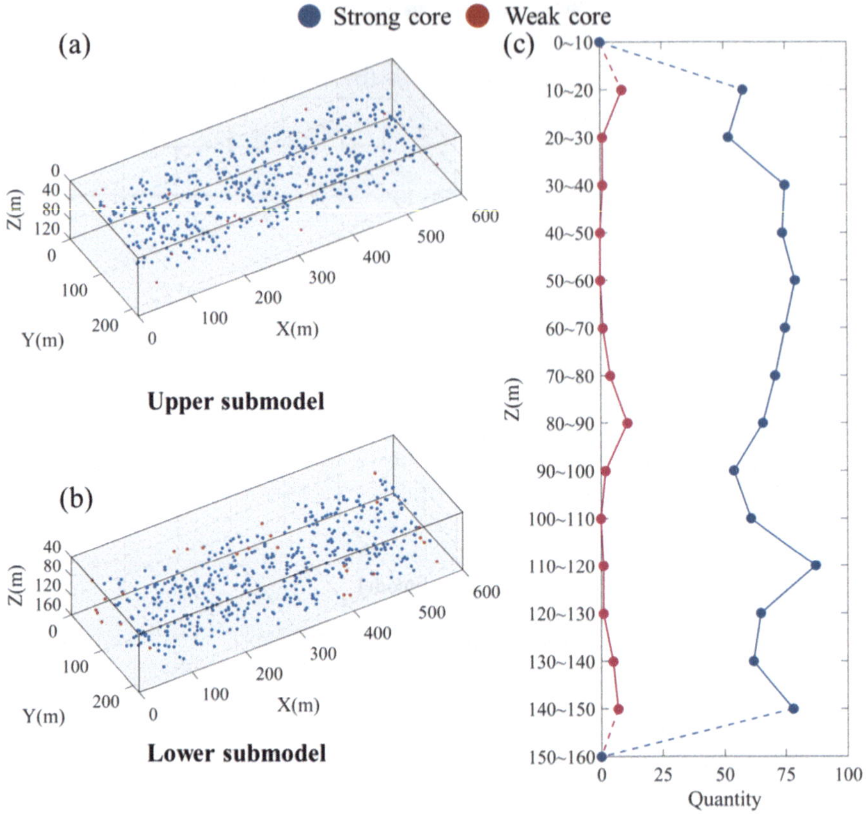

Fig. 8.8 Source clustering results. **a** Sources in the upper submodel; **b** sources in the lower submodel; **c** histograms showing source counts for each submodel

the JP underground laboratory exerts only a slight effect on the inversion results. In the depth range corresponding to the laboratory caverns (60–110 m), minor low-velocity anomalies can be seen, as marked by the red circle in Fig. 8.10b. The positions of these anomalies correspond closely with the actual sites of the excavated caverns.

In the depth ranges of 10–40 m and 120–150 m, noticeable variations in the *CC* are observed, with the inversion again performing better for low-velocity anomalies than for high-velocity anomalies within the checkerboard pattern. The inversion accuracy is highest in the 40–120 m depth range, exhibiting a trend consistent with that shown in Fig. 8.9c. The average *CC* within this interval reaches 0.79, representing only a marginal reduction of 0.05 compared to the Model-1 result. These findings confirm that the presence of excavated voids does not significantly impair the performance of the SDD method.

In the absence of underground structures, seismic waves propagate unobstructed through the rock mass. Conversely, excavated regions obstruct direct wave transmission, leading to more complex propagation paths before the waves reach the sensors.

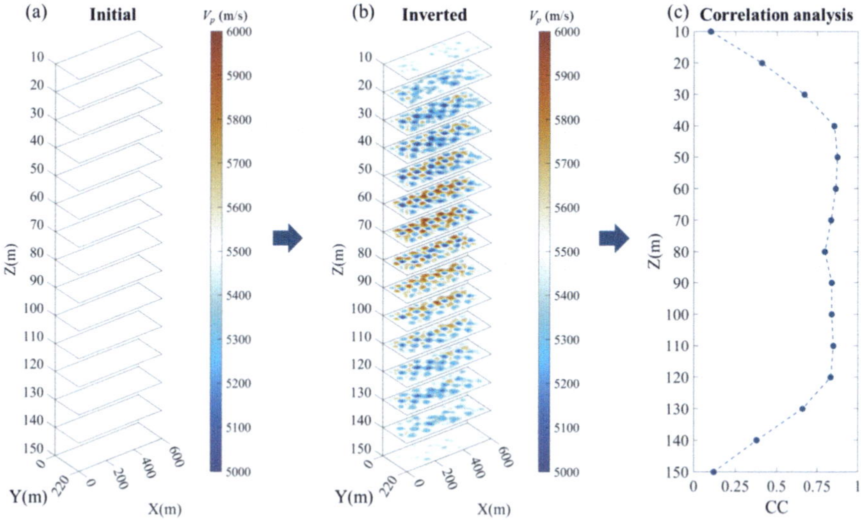

Fig. 8.9 Checkerboard inversion excluding the Jinping laboratory structures. **a** Initial velocity model; **b** final inverted velocity field; **c** *CC* analysis of the initial and inverted models

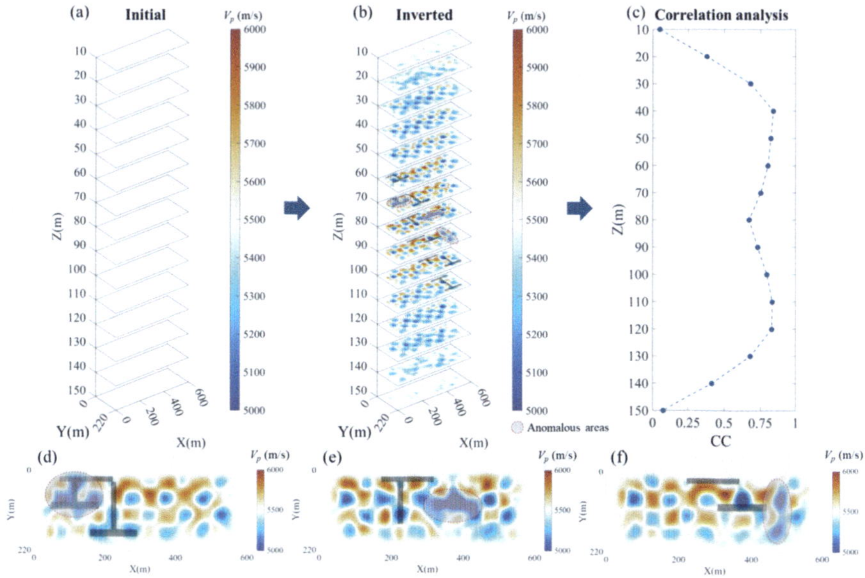

Fig. 8.10 Checkerboard inversion incorporating Jinping underground structures. **a** Initial velocity model; **b** inverted velocity field; **c** cross-correlation between the initial and inverted models; **d** enlarged section at $Z = 70$ m from (**b**); **e** enlarged section at $Z = 80$ m; **f** enlarged section at $Z = 90$ m

This distinction accounts for the slightly reduced inversion accuracy observed in Model-2 relative to Model-1. Overall, the checkerboard tests demonstrate that the SDD method achieves reliable performance at a resolution of approximately 50 m. The minor deviations observed, together with the generally high inversion accuracy, confirm the practical applicability of this method to the Jinping project.

8.5 Realistic Synthetic Dynamic Inversion Tests

8.5.1 Design of the Excavation Process

Two static checkerboard tests based on the Jinping MS monitoring configuration have demonstrated the feasibility of the SDD method for application in the Jinping project. This section further investigates the dynamic inversion capability of the SDD method during cavern excavation.

The four laboratory halls are of identical dimensions. To facilitate direct observation of the velocity field inversion results within the central region of the model, only the two centrally located halls B and C are selected for simulated excavation. Halls A and D are excluded from the excavation sequence to avoid excessive complexity and to simplify the analysis of the test outcomes.

The excavation process is divided into eight sequential steps, as illustrated in Fig. 8.11a. Steps 1 through 4 correspond to the excavation of hall B, while steps 5 through 8 are dedicated to hall C. Both halls are excavated using the same strategy: advancing from the central axis outward toward both sidewalls, with each step progressing 32.5 m. Accordingly, a total of eight velocity field inversions are performed, one following each excavation step.

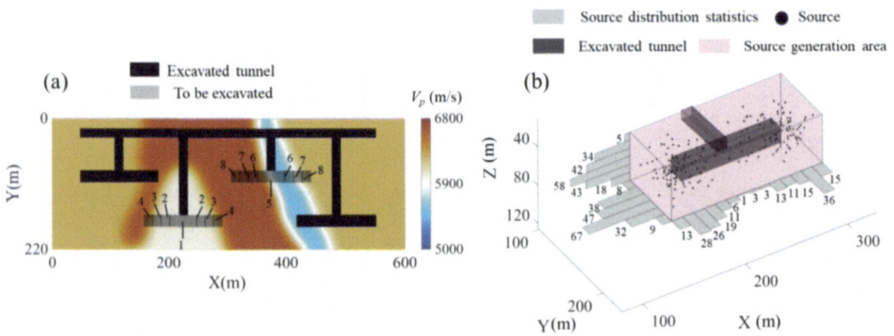

Fig. 8.11 Parameters for dynamic inversion. **a** True model velocity slice; **b** schematic showing the generation of MS sources

8.5.2 Set-Up for the Dynamic Test

(1) Real Model

Following Guo (2018), a simplified yet realistic velocity model of the Jinping underground laboratory is constructed, as illustrated in Fig. 8.11a. A distinct low-velocity zone is observable near $X = 400$ m, while a broad high-velocity region dominates the interval between 180 m and 400 m along the X-axis, within which an intermediate-velocity zone is embedded. Overall, the velocity model comprises four distinct zones characterized by velocities of 5500, 6000, 6300, and 6700 m/s.

(2) Synthetic MS Sources

Research indicates that a significant number of MS events are induced in rock masses during or following excavation activities (Lai et al. 2022). Accordingly, MS sources are generated after each excavation step, with the corresponding quantities listed in Table 8.2.

After each excavation step, the spatial distribution of MS events close to the excavation face shows a regular pattern (Wang et al. 2025). In this simulation, a probability function dependent on distance is utilized for source generation, making event density reduce as the distance from the current working faces increases. Only events occurring within a certain radius of the tunnel boundaries are taken into account, following practical MS monitoring limitations (Wang et al. 2025). Figure 8.11b displays this spatial distribution for excavation step 3, with allowable source regions (except the black-colored excavated areas) presented in pink. Gray histograms quantifying the number of MS events along the x-, y-, and z-axes are superimposed on the 3D spatial map. The statistical results verify that the generated sources meet the specified criteria.

(3) Add Errors in the Absolute P-Wave Arrival Time Picking Manually

Given that manually or automatically picked arrival times in practical engineering are rarely error-free, artificial random perturbations within a range of ± 1 ms are introduced into all theoretical arrival times. These perturbed traveltime data are then employed for source localization and subsequently used as input for velocity inversion.

(4) Parameter Settings

In contrast to idealized checkerboard tests, the absence of prior knowledge regarding P-wave velocity and the presence of arrival time errors in this chapter necessitate the implementation of P-wave estimation and data quality control procedures. Moreover,

Table 8.2 Quantity of MS sources generated after each stage

Step	1	2	3	4	5	6	7	8
Quantity	100	150	200	250	100	150	200	250

Table 8.3 Source clustering parameters for dynamic inversion

Parameter	Clustering					Data quality controlling
	α (m)	β	R_{dmax} (m)	Num_{smin}	Num_{smax}	σ_r (m)
Value	20	10	50	8	20	40

owing to differences in the spatial scale of the inversion domain and the distribution characteristics of the microseismic sources, the clustering process requires adjusted parameters. The specific parameters employed for data quality control and clustering are summarized in Table 8.3.

8.5.3 Velocity Model Inversion with Biased Data

This part demonstrates the inversion process under perturbed data conditions, taking excavation stage 1 as an example. As depicted in Fig. 8.12, 100 sources are created according to the procedure outlined in Sect. 8.5.2. Artificial random errors are introduced to the precise arrival times acquired through the MSFM. As all MS events are simulated without corresponding waveforms, the P-wave arrival picking process is not elaborated here.

First, to perform inversion, an estimate of the P-wave velocity is needed. An active source at coordinates (230, 170, 60) is activated and detected by all sensors. Traveltimes are calculated using Eq. 8.1 from the recorded absolute arrival times and the known source triggering time. Figure 8.13a shows source-receiver distances versus corresponding traveltimes; linear regression gives an estimated P-wave velocity of 5892.3 m/s.

Using a uniform velocity of 5,892.3 m/s, the conventional location method (Feng et al. 2017) is applied to calculate source positions and initiation times. A 40-sensor

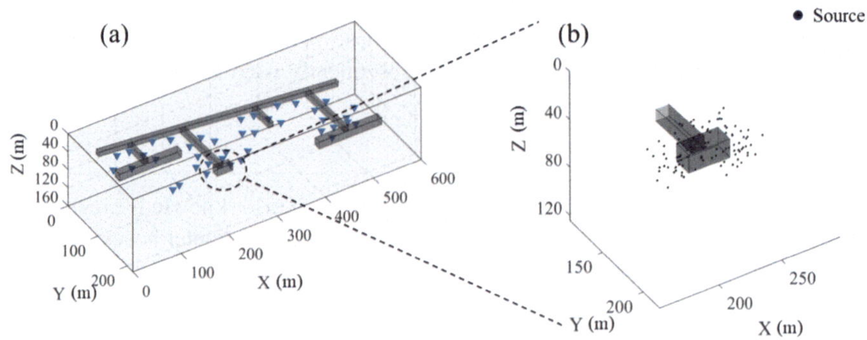

Fig. 8.12 Excavation step 1 model configuration **a** three-dimensional representation of the entire JP laboratory; **b** magnified view of hall B

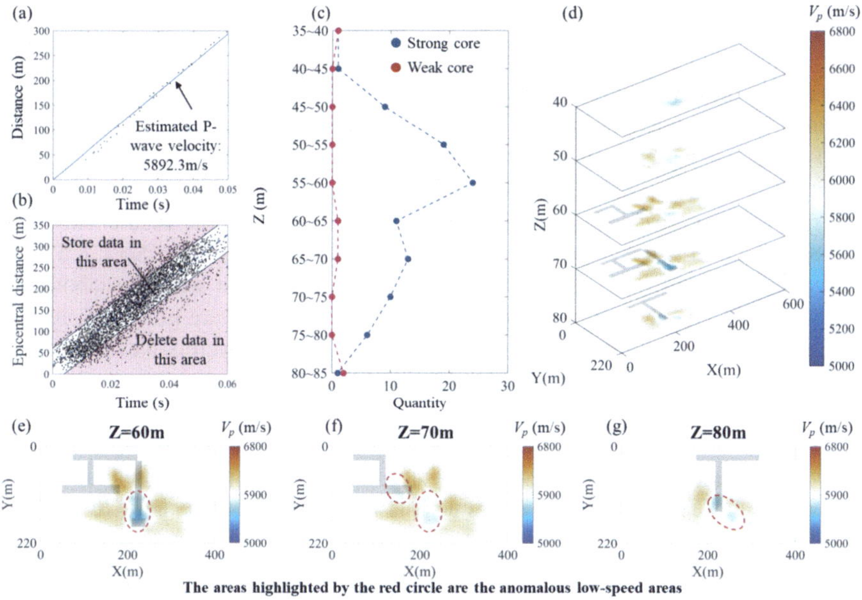

Fig. 8.13 Post-excavation step 1 inversion details. **a** P-wave velocity estimate from active source data; **b** outlier removal during quality control; **c** source distribution statistics. **e–g** Enlarged views of the inverted field at depths of 60, 70, and 80 m

array captures 4000 arrival times (100 events × 40 sensors), with each time being artificially perturbed by noise. The obtained source positions produce epicentral distances d and traveltimes t, as illustrated in Fig. 8.13b. Data filtering keeps only those events within the limits set by two regression lines (Eq. 8.2), using an intercept offset threshold $o_r = 40$ (as indicated in Table 8.3). Figure 8.13c shows the clustering outcomes, recognizing 95 strong—core sources and 5 weak—core sources among the 100 MS events. The starting velocity for inversion is fixed at 5,900 m/s, close to the estimated value of 5,892.3 m/s.

Figure 8.13d displays the inversion results following excavation step 1. Enlarged views of the depth interval $Z = 60$–80 m (the depth range of the Jinping laboratory) within Fig. 8.13d are shown in Fig. 8.13e–g. Anomalous low-velocity areas, highlighted by red circles, are present in these figures, features that do not exist in the true velocity model. These anomalies exhibit strong spatial correspondence with the locations of the laboratory structures and are therefore attributed to their influence.

New MS events are subsequently generated based on the inverted velocity model, and the inversion results from the current step serve as the initial model for the next excavation stage. In this manner, the workflow establishes an iterative updating process for the velocity field.

8.5.4 Results and Analysis

Figure 8.14 presents the dynamic inversion results of the velocity field during the sequential excavation of laboratory halls B and C. The inversion domain spans depths of 40–90 m for steps 1–4 (Hall B) and 60–100 m for steps 5–8 (Hall C).

As the inversion iterations progress, the SDD method progressively reconstructs the true velocity structure. Figure 8.14a, corresponding to the result shown in Fig. 8.13d, exhibits the anomalous low-velocity zones discussed in Sect. 8.5.3. In subsequent steps (Fig. 8.14b–d), these anomalies, associated with the laboratory structures and visible as blue regions at depths of 70–80 m, are gradually attenuated, while the surrounding velocity field is successfully recovered. Nevertheless, a small portion of the velocity field remains unreconstructed due to the presence of the cavern.

Figure 8.14e–h depict the inversion procedure commencing around Hall C. In the depth interval of 70–90 m, a vertically aligned low-velocity zone, centered close to $X = 400$ m (existent in the true model), is precisely reconstructed. Conversely, at a depth of 60 m, which is shallower, the anomalous region encircled in red in Fig. 8.14e gradually decreases in size with iterative operations. Importantly, the inversion quality above the cavern surpasses that below it.

Although the method exhibits a strong capability in distinguishing between low-velocity (5500–6000 m/s) and high-velocity (> 6300 m/s) regions, its resolution diminishes when differentiating velocities in the intermediate range (6000–6300 m/s) from adjacent high-velocity zones. *CC* analysis conducted after each excavation step reveals consistently higher values within the 60–80 m depth interval compared to other depths, with a maximum *CC* of 0.74 (Fig. 8.15a). Lateral comparison across depth ranges shows stable *CC* values during steps 6–8, indicating the robustness of the SDD method.

Error analysis, performed by calculating the difference between the inverted and true velocity fields after each step, reveals two distinct evolutionary patterns (Fig. 8.15b, c). First, the standard deviation and mean error remain relatively stable at approximately 154 m/s and 212 m/s, respectively, both exhibiting gradual downward trends throughout the eight inversion cycles. Second, a transient error surge ($\approx$ 1350 m/s) occurs during the fifth inversion step, followed by a reduction to approximately 950 m/s in subsequent iterations. The pink-shaded regions in Fig. 8.15b represent the proportional distribution of specific error ranges, with the width of each shaded area reflecting the corresponding proportion. Approximately 88.7% of inversion errors fall within 0–600 m/s across all stages, and the shapes of the shaded regions remain broadly consistent.

The error surge observed during Step 5, coinciding with the onset of Hall C excavation, is closely related to the spatial configuration of the velocity field and the cavern location. Hall C lies at the boundary between the highest- and lowest-velocity zones, a feature that becomes particularly pronounced in Step 5 (Fig. 8.11a). Moreover, according to the source-generation criteria adopted in this study, a higher density of seismic sources is generated near the tunnel face and sidewalls. This results in a

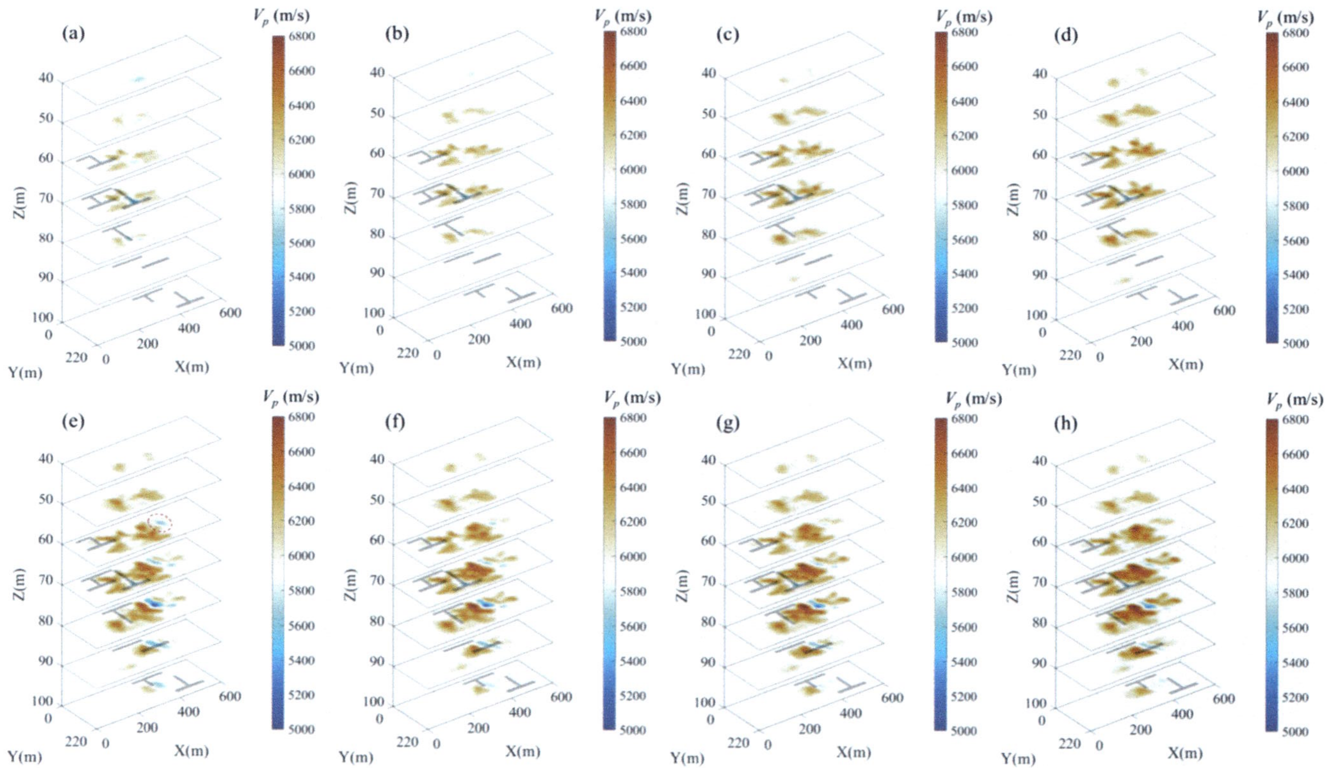

Fig. 8.14 Sequential inversion outcomes for excavation steps 1–8. **a** Step 1; **b** Step 2; **c** Step 3; **d** Step 4; **e** Step 5; **f** Step 6; **g** Step 7; **h** Step 8

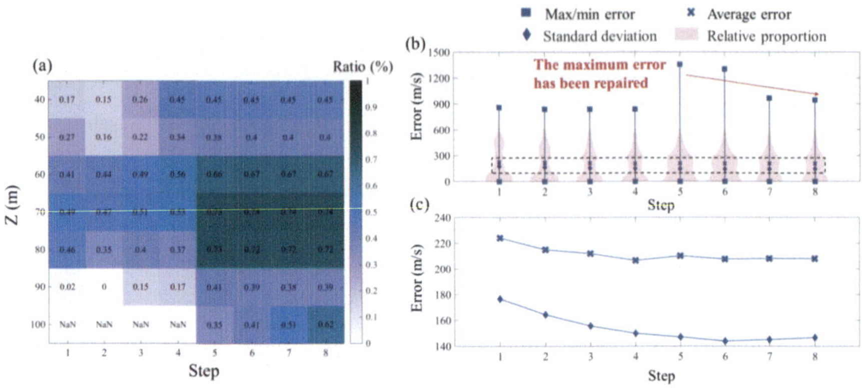

Fig. 8.15 Evaluation of dynamic inversion accuracy. **a** *CC* results; **b** error map of the inverted velocity field; **c** enlarged section corresponding to the black box in (**b**)

greater number of sources located in the transitional zone between high- and low-velocity regions during Step 5 compared to Steps 6–8, potentially leading to misclassification between velocity domains. The magnitude of the observed error ($\approx$ 1300 m/s) aligns closely with the velocity contrast across this boundary, suggesting that this is the primary cause of the transient surge.

Overall, these results demonstrate that the SDD method maintains stable performance in dynamic inversion tasks, even when the initial input data contain errors.

8.6 Discussion

This research assesses the resolution and dynamic inversion capabilities of the SDD method via two checkerboard experiments and dynamic inversion tests rooted in the Jinping MS monitoring configuration. Several aspects need further exploration:

(1) Stratification considerations and impacts of no stratified inversion

The proposed SDD method explicitly requires the division of the model into upper and lower submodels during the stratification and clustering step. It should be noted that sources located exactly at the dividing depth of $Z = 80$ m do not participate in either submodel unless manually assigned to one of them. For sources precisely at this boundary, manual assignment to either the upper or lower submodel is permitted.

The overlapping region between the two submodels leads to dual inversion of the velocity field within that depth interval. In practice, only the velocity fields at the stratification boundary (indicated by the red line in Fig. 8.7) are retained, while those at other depths are discarded. As illustrated in Fig. 8.7, depths from 40 m to 80 m are included in both submodel inversions. Therefore, the results from the upper submodel for depths greater than 80 m and from the lower submodel for depths less than 80 m should be excluded. At the boundary depth of $Z = 80$ m, the velocity field is obtained by weighted averaging based on the number of sources contributing from each submodel.

Omitting the stratification step leads to a significant degradation in inversion accuracy. The *CC* of the nonstratified inversion result (Fig. 8.16b) is 0.14 lower than that of the stratified result at the same depth (Fig. 8.16a $Z = 90$ m), demonstrating poor performance. This comparison underscores the critical importance of stratification for the effective application of the SDD method.

(2) The Impact of Errors in Input Data on Inversion

The absolute arrival times of P-waves constitute the fundamental input data for the SDD method. In addition to algorithmic performance, the accuracy of these data represents a controllable factor in practical applications. When random errors of up to ±1.2 ms are introduced into the ideal checkerboard data, the *CC* decreases by only 0.03 compared to the result obtained with ideal data (Fig. 8.16a). These results demonstrate that the SDD method exhibits satisfactory robustness to errors in the input data. Notably, current P-wave picking technology for microseismic monitoring (Xiao et al. 2025) achieves an accuracy of approximately ±1 ms, which meets the data precision requirements of the proposed method.

(3) The Sensitivity of Stratification Depth

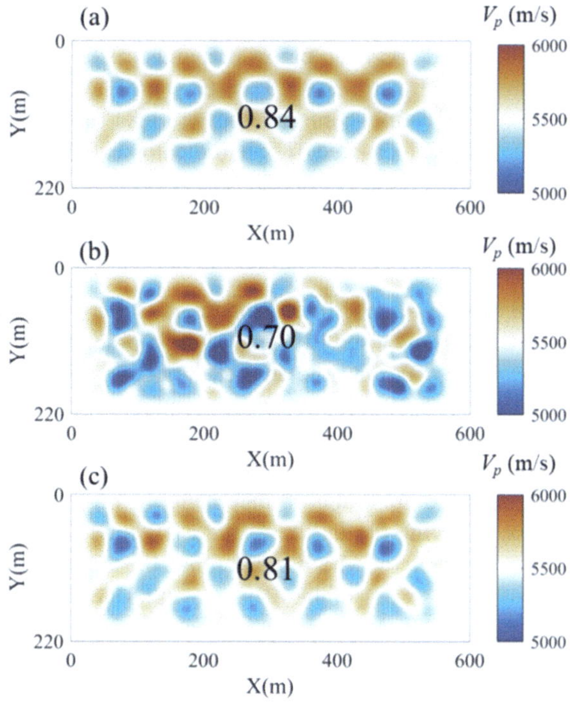

Fig. 8.16 Checkerboard tests under different conditions. **a** Ideal input data scenario; **b** inversion without applying stratification; **c** inversion result with erroneous input data

In this study, a fixed stratification depth of $Z = 80$ m is adopted for all tests, determined based on the mean depth of the sensors. The objective is to ensure that the depth ranges over which sensors and sources overlap are evenly distributed between the upper and lower submodels. In deep underground engineering contexts, the irregular spatial distribution of sources and sensors introduces instability into double-difference tomography. To mitigate the adverse effects of this instability, it is essential to balance the overlapping depth intervals. Accordingly, the mean sensor depth is selected as the stratification plane.

To assess the sensitivity of the inversion results to the choice of stratification depth, two additional tests are conducted using the checkerboard model, with stratification depths set to 70 m (Fig. 8.17b) and 60 m (Fig. 8.17a). The baseline result with a stratification depth of 80 m corresponds to that shown in Fig. 8.9. Varying the stratification depth alters the extents of the upper and lower submodels, thereby affecting the distribution of sensor–source overlap. For comparison, the velocity slice at $Z = 80$ m, which lies within the lower submodel for the 70 m and 60 m cases, is selected for analysis.

When the stratification depth is decreased from 80 m to 70 m, the resulting velocity field exhibits minimal change, with the *CC* decreasing by only 0.02. However, when the depth is further reduced to 60 m, the *CC* drops sharply to 0.72. These results indicate that deviations of approximately 10 m from the mean sensor depth have a limited impact on inversion accuracy. In contrast, a deviation of 20 m leads to a substantial degradation in performance.

(4) Limitations

The suggested approach places demands on the arrangement of sensors. For inversion to succeed, a crisscrossed sensor-source layout in the horizontal plane is needed; otherwise, a distinct configuration will lead to method failure. The extent of interlacing can be measured by an indicator called the *Mix*, which is computed as follows:

$$\text{Mix} = \frac{\text{mix}_{\text{area}}}{\min\{\text{area1}, \text{area2}\}} \tag{8.18}$$

where Mix represents the mixing degree. area1 and area2 stand for the areas of the quadrilaterals enclosed by the outermost points of the sensors and sources, respectively. mix_{area} signifies the area of their overlapping part. The outermost points are described as the four points corresponding to the minimum and maximum coordinates in both horizontal directions. Fig. 8.18 shows four different distributions with a decreasing *Mix*. The distributions in Fig. 8.18a, b are appropriate for the SDD method. In contrast, the arrangements in Fig. 8.18c, d cannot produce inversion results, implying the failure of the SDD method. This indicates that SDD is only applicable when the Mix is greater than 0.28. The dense monitoring network around the excavated section of the JP underground laboratory meets the required interleaved sensor-source distribution and is therefore suitable for this method.

This work presents studies, along with existing research on double-difference tomography (Guo and Thurber 2021), all of which are based on isotropic media.

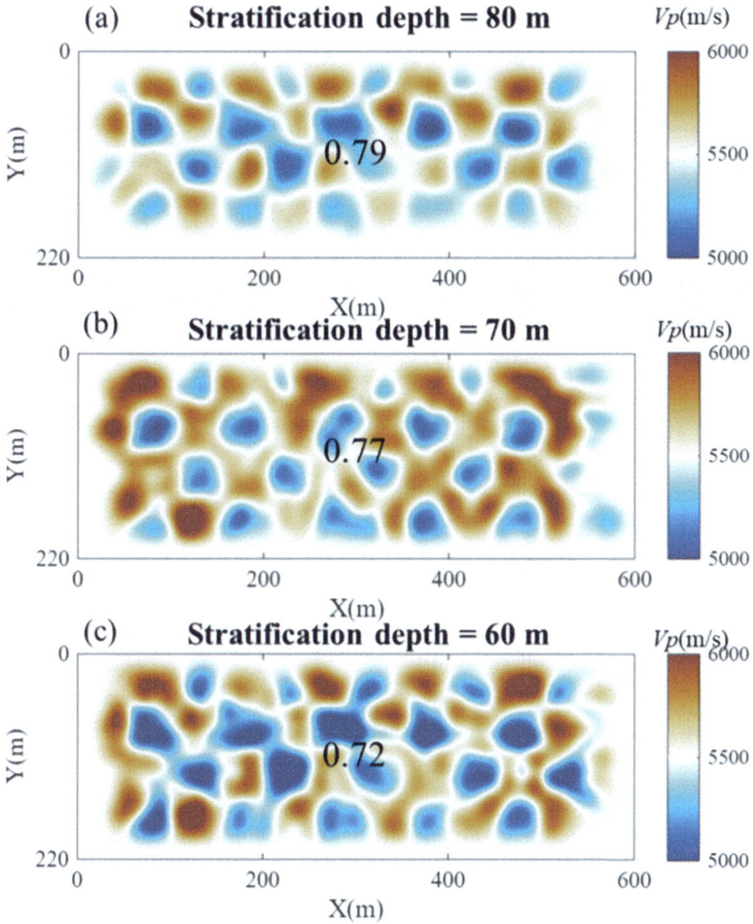

Fig. 8.17 Inversion results of different stratification depths. **a** $Z = 80$ m. **b** $Z = 70$ m. **c** $Z = 60$ m

However, certain limitations remain when applying these methods to other media types.

In anisotropic media, the velocity of wave propagation and ray paths rely on direction (Cerveny and Firbas 1984; Sadri and Riahi 2010). Ignoring this anisotropy may cause substantial inversion errors. In strongly heterogeneous media, waves experience intricate refraction and reflection (Gu and Jing 2020), and when the scale of velocity anomalies is equal to or less than the wavelength, conventional ray-tracing methods fail (Yang and Hung 2005). In multiple scattering media, incident waves interact with heterogeneities to produce a great number of scattered waves that convey valuable information regarding internal structures (Williamson 1991). These scattered parts are completely neglected by single-ray-path-based approaches, thus leading to inaccurate parameter inversion.

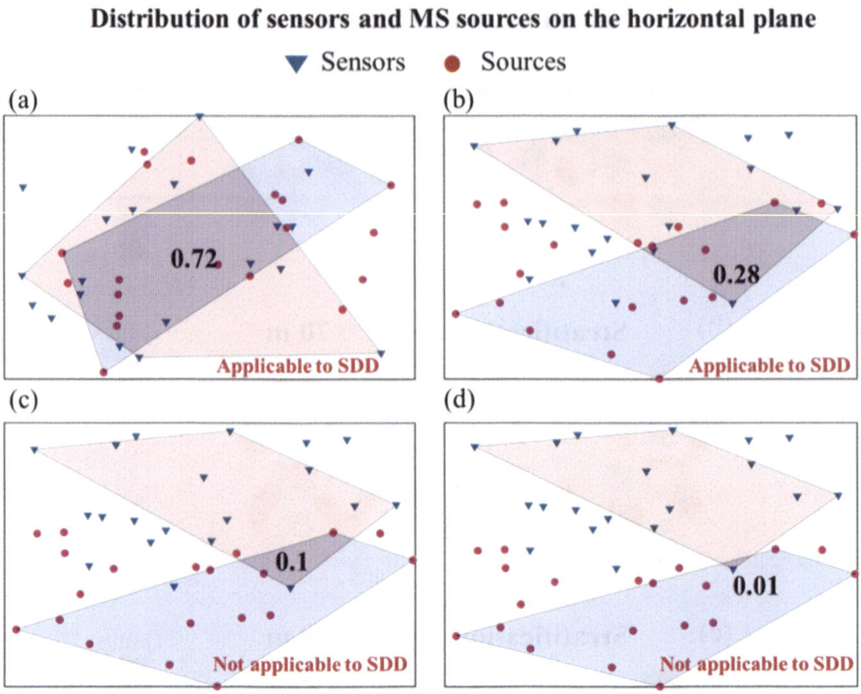

Fig. 8.18 Plan-view illustrations of sensor–source layouts with Mix values: **a** 0.72, **b** 0.28, **c** 0.1, and **d** 0.01

Although double-difference tomography has been widely and successfully applied to inversion in inhomogeneous media (Got et al. 2006, 2008), its extension to anisotropic media or incorporation of attenuation effects remains unexplored due to high computational complexity. Given the advances already made in ray theory for anisotropic media, integrating these developments into the double-difference framework may offer a promising direction for future research.

8.7 Conclusion

This study proposes a stratified double-difference (SDD) tomography method that enables dynamic inversion of three-dimensional velocity fields in large underground caverns using MS events. The methodological framework consists of four main components: data preprocessing, quality control, stratification and clustering, and velocity field inversion. Based on the monitoring configuration of the Jinping project, the method is validated through two checkerboard tests and a dynamic inversion simulating the sequential excavation of halls B and C. The performance of the

SDD method is quantitatively evaluated using the CC between the inverted and true velocity models. The following conclusions are drawn.

(1) Two idealized checkerboard tests based on the Jinping monitoring setup, one incorporating excavated laboratory structures and the other excluding them, demonstrate that the SDD method effectively distinguishes between high- and low-velocity zones. The achievable resolution is approximately 50 m. The average *CC* for the test including underground structures reaches 0.79, only 0.05 lower than that of the test without such structures, indicating that the presence of excavated regions has minimal interference with the inversion performance.
(2) A simulation of the excavation processes in halls B and C of the Jinping project is conducted using 1,400 synthetic sources generated according to realistic spatiotemporal MS distributions. Random errors of up to ± 1 ms are introduced into the absolute arrival times to reflect practical conditions. As excavation progresses, the surrounding velocity structures are progressively reconstructed. An inclined low-velocity zone adjacent to hall C is clearly resolved, and an average *CC* of 0.71 confirms the consistency between the inversion results and the true model. These findings demonstrate that the SDD method performs reliably in dynamic inversion scenarios.
(3) The SDD method exhibits tolerance to errors in input arrival time picks. Checkerboard tests based on the JP configuration show that the *CC* decreases by only 0.03 when ± 1 ms errors are introduced. Furthermore, the implementation of the stratification strategy improves the CC from 0.70 to 0.84, indicating that this approach effectively mitigates accuracy degradation caused by the spatial distribution of MS events and sensors.

References

Bai C, Greenhalgh S, Zhou B (2007) 3D ray tracing using a modified shortest-path method. Geophysics 72(4):27–36

Bunks C, Saleck FM, Zaleski S, Chavent G (1995) Multiscale seismic waveform inversion. Geophysics 60(5):1457–1473.

Cerveny V, Firbas P (1984) Numerical modelling and inversion of travel times of seismic body waves in inhomogeneous anisotropic media. Geophys J Int 76(1):41–51

Chen J, Wu SC, Xu MJ, Nagaso M, Yao JY, Wang K, Li TJ, Bai YM, Tong P (2023) Adjoint-state teleseismic traveltime tomography: method and application to Thailand in Indochina Peninsula. J Geophys Res: SolidEarth 128(12):e2023JB027348

Deng D (2020) DBSCAN clustering algorithm based on density. In: 2020 7th international forum on electrical engineering and automation (IFEEA). IEEE, pp 949–953

Dong LJ, Pei ZW, Xie X, Zhang YH, Yan XH (2023) Early identification of abnormal regions in rock-mass using traveltime tomography. Engineering 22:191–200

Feng GL, Feng XT, Chen BR, Xiao YX (2017) Performance and feasibility analysis of two micro-seismic location methods used in tunnel engineering. Tunnell Underground Space Technol 63:183–193

Gu J, Jing Y (2020) A modified mixed domain method for modeling acoustic wave propagation in strongly heterogeneous media. J Acoust Soc Am 147(6):4055–4068

Guo HS (2018) Quantitative evaluation method for the integrity and fracture evolution of surrounding rock of deep hard rock tunnel. DongBei University

Guo H, Thurber C (2021) Double-difference seismic attenuation tomography method and its application to The Geysers Geothermal Field. California. Geophys J Int 225(2):926–949

Got J, Monteiller V, Virieux J, Okubo P (2006) Estimating crustal heterogeneity from double-difference tomography. Pure Appl Geophys 163(2–3):405–430

Got J, Monteiller V, Virieux J, Operto P (2008) Potential and limits of double-difference tomographic methods. Geophys Prospect 56(4):477–491

Haralick RM, Linda GS (1992) Computer and robot vision. II, Addison-Wesley: 316-317.

Hassouna MS, Farag AA (2007) Multistencils fast marching methods: a highly accurate solution to the eikonal equation on cartesian domains. IEEE Trans Pattern Anal Mach Intell 29(9):1563–1574

Jia BX, Li F, Zhou LL, Wang S, Liu JS (2021) Joint arrival-time picking method of microseismic P-wave and S-wave based on time-frequency analysis. Rock Soil Mech 42(5):4

Jiang P, Cao S, Guo S, Ren Y, Li Y (2024) Deep learning based optimization framework for full-waveform inversion in Tunnels. IEEE Trans Geosc Remote Sens 62:1–11

Junho U, Clifford T (1987) A fast algorithm for two-point seismic ray tracing. Bull Seismol Soc Am 77(3):972–986

Koketsu K, Sekine S (1998) Pseudo-bending method for three-dimensional seismic ray tracing in a spherical earth with discontinuities. Geophys J Int 132(2):339–346

Kasdorf S, Troksa B, Key C, Harmon J, Branislav MN (2021) Advancing accuracy of shooting and bouncing rays method for ray-tracing propagation modeling based on novel approaches to ray cone angle calculation. IEEE Trans Antennas Propag 69(8):4808–4815

Lai XP, Jia C, Cui F, Chen JQ, Zhou YP, Feng GG, Gao YJ (2022) Microseismic energy distribution and impact risk analysis of complex heterogeneous spatial evolution of extra-thick layered strata. Sci Rep 12(1):10832

Li JC, Li HB, Ma GW, Zhou YX (2013) Assessment of underground tunnel stability to adjacent tunnel explosion. Tunnell Underground Space Technol 35:227–234

Liu B, Zhang F, Li S, Liu L, Li Y, Xu X, Liu Z, Zhang Q (2019) Full waveform inversion based on inequality constraint for cross-hole radar. J Appl Geophys 162:118–127

Liu QS, Wu J, Zhang XP, Tang LX, Bi C, Li WW, Xu JL (2020) Microseismic monitoring to characterize structure-type rockbursts: a case study of a TBM-excavated tunnel. Rock Mech Rock Eng 53:2995–3013

Liu L, Li S, Xiao Y, Chen S, Wang Z, Feng G, Wang Y (2023) An inverted heterogeneous velocity model for microseismic source location in deep buried tunnels. Rock Mech Rock Eng 56(7):4855–4880

Lin Y, Theiler J, Wohlberg B (2023) Physics-guided data-driven seismic inversion: Recent progress and future opportunities in full-waveform inversion. IEEE Signal Process Magazine 40(1):115–133

Lin CJ, Li SC (2014) Tunnel seismic prediction (TSP) and its application in tunnel engineering. Appl Mech Mater 501:1779–1782

Lu X, Wang W, Yang C, Hu X, Liao X, Fu Z (2024) Exploring the influence of seismic source and improvement methods on tunnel seismic prediction. Lithosphere 2.

Musayev K, Lamert A, Hackl K, Friederich W, Baitsch M (2016) Predicting the geological structure ahead of a tunnel using full waveform inversion—A blind test. In: Proceedings of the EAGE conference and exhibition, Vienna, Austria

Pratt RG (1990) Frequency-domain elastic wave modeling by finite differences: a tool for crosshole seismic imaging. Geophysics 55(5):626–632

Qu XX, Liu SX, Wang F (2014) A new ray tracing technique for crosshole radar traveltime tomography based on multistencils fast marching method and the steepest descend method. In: Proceedings of the 15th international conference on ground penetrating radar. IEEE, pp 503–508

Sadri M, Riahi MA (2010) Ray tracing and amplitude calculation in anisotropic layered media. Geophys J Int 180(3):1170–1180

Sethian JA, Popovici AM (1999) 3-D traveltime computation using the fast marching method. Geophysics 64(2):516–523

Shih RC, Chen W (1996) Iterative pre-stack depth migration with velocity analysis. Terrestrial Atmos Oceanic Sci 7(2):149–158

Waldhauser F, Ellsworth WL (2000) A double-difference earthquake location algorithm: method and application to the northern Hayward fault, California. Bull Seismol Soc America 90(6):1353–1368

Wang J, Zuo YJ, Dong LJ, Yan XH (2025) Spatiotemporal clustering of microseismic signals in mining areas: a case study of the Baoji lead-zinc mine in Shaanxi, China. Eng Geol 352:108057

Williamson PR (1991) A guide to the limits of resolution imposed by scattering in ray tomography. Geophysics 56(2):202–207

Wu S, Chen J, Tong P (2025) Seismic azimuthal anisotropy of New Zealand revealed by adjoint-state traveltime tomography. Earth Planet Sci Lett 660:119362

Wolfe CJ (2002) On the mathematics of using difference operators to relocate earthquakes. Bull Seismol Soc Am 92(8):2879–2892

Xiao Y, Chen S, Wang Z, Liu L, Du C (2023) Analyzing the stability of rock surrounding deep cross-tunnels using a dynamic velocity field. Sustainability 15(20):15139

Xiao YX, Chen SJ, Feng GL, Liu L, Li SJ, Zhou JB, Yan TZ, Guo JN (2025) Picking P-wave arrival times from microseismic signals in tunnels with low signal-to-noise ratios without noise filtering. Soil Dyn Earthquake Eng 192:109288

Xue Y, Li Z, Qiu D, Yang W, Zhang L, Tao Y, Zhang K (2019) Prediction model for subway tunnel collapse risk based on delphi-ideal point method and geological forecast. Soil Mech Found Eng 56(3):191–199

Yang H, Hung S (2005) Validation of ray and wave theoretical travel times in heterogeneous random media. Geophys Res Lett 32(20)

Yao ZS, Roberts RG, Tryggvason A (1999) Calculating resolution and covariance matrices for seismic tomography with the LSQR method. Geophys J Int 138(3):886–894

Yu M, Cheng F, Liu J, Peng D, Tian Z (2022) Frequency-domain full-waveform inversion based on tunnel space seismic data. Engineering 18(11):207–217

Zhang H, Thurber C (2003) Double-difference tomography: The method and its application to the Hayward Fault, California. Bull Seismol Soc Am 93(5):1875–1889

Zhang H, Thurber C (2006) Development and applications of double-difference seismic tomography. Pure Appl Geophys 163(2):373–403

Zhang H, Thurber C, SatoshiIde D, Gregory C, Beroza A (2004) High-resolution subducting-slab structure beneath northern Honshu, Japan, revealed by double-difference tomography. Geology 32(4):361–364

Xiao Y, Chen S, Liu L et al (2024) Determining S-wave arrival times for microseismic monitoring in tunnels based on instantaneous phase difference of sensors[J]. Rock Mech Rock Eng 57(3):1781–1797

Chapter 9
Conclusion

Chapters 2–8 present a systematic framework for processing MS monitoring data in deep underground tunnels. The chapters address key stages of MS data analysis, including signal identification, phase picking, event localization, velocity inversion, and stability assessment of surrounding rock. The proposed methods are designed specifically for the complex and noisy environment of deep tunnel construction, where accurate interpretation of MS signals is essential for understanding rock fracturing processes and improving rockburst prediction.

9.1 Automated Identification of Rock Fracture Signals (Chap. 2)

The first step in MS data processing is distinguishing rock fracture signals from numerous noise signals generated during construction activities such as drilling, blasting, machinery operation, and vehicle movement. Manual signal identification has traditionally been the dominant approach in tunnel monitoring; however, it is labor intensive and inefficient when large volumes of waveform data are collected.

Chapter 2 proposes an automated signal identification method that digitizes the manual interpretation process. Instead of relying on large training datasets such as deep learning models, the method extracts waveform characteristics that mimic human judgment when visually inspecting MS signals. The identification procedure consists of three major stages:

1. Normalization and Clustering Analysis

The waveform amplitude is normalized and analyzed using a K-means clustering algorithm to determine whether a signal contains distinct pulse characteristics typical of rock fracturing.

2. Waveform Extremum Extraction

© Tongji University Press Co., Ltd 2026

Y. Xiao et al., *Near-Field Microseismic Data Processing in Deep Tunnels*,
https://doi.org/10.1007/978-981-92-3208-6_9

The signal is modified to highlight local extrema, and the algorithm evaluates whether the waveform contains a single pulse or multiple pulses. Rock fracture signals typically exhibit a single dominant pulse, whereas construction noise often produces multiple pulses.

3. Key Point Extraction and Waveform Shape Analysis

Critical points around the pulse peak are identified through interpolation. A ratio parameter describing the waveform shape is then calculated to verify whether the signal matches the typical pattern of rock fracture events.

Two thresholds control the classification process: one describing waveform clustering characteristics and the other representing pulse shape constraints. These thresholds are determined from manually labeled datasets using evaluation metrics such as accuracy rate and miss rate.

The method was validated using MS monitoring data from two tunnels excavated with different construction techniques. In a tunnel bored by a tunnel boring machine (TBM), the method achieved an average identification accuracy of approximately 95.6% with a false-negative rate of approximately 4%. In a drill-and-blast tunnel, the accuracy remained high at approximately 93.9%. The results demonstrate that the digitalization approach can reliably identify rock fracture signals with relatively small training datasets and maintain stable performance under different construction conditions.

9.2 P-Wave Arrival Time Picking Under Low Signal-to-Noise Ratios (Chap. 3)

Accurate determination of P-wave arrival times is fundamental for locating MS events. Conventional methods such as the STA/LTA algorithm and the AIC method perform well in high SNR conditions but become unreliable when noise levels are high. In tunnel environments, MS signals frequently suffer from severe noise contamination, making reliable arrival time picking difficult.

Chapter 3 introduces a new method called IPDI_P, which identifies P-wave arrival times directly from raw signals without noise filtering. The method is based on the observation that signals generated by the same MS event are recorded at multiple sensors with slightly different travel times. When the propagation path difference between two sensors is smaller than one P-wave wavelength, a condition commonly satisfied in tunnel monitoring, the phase difference between the signals exhibits a predictable behavior.

The IPDI_P approach computes the instantaneous phase of two signals using the Hilbert transform and evaluates their phase difference over time. A characteristic peak appears in the phase difference intensity function when the P-wave reaches a sensor. The arrival time is therefore determined by identifying the first significant peak in this function.

The method includes four main steps: signal labeling, auxiliary signal selection, phase difference computation, and peak detection for arrival picking. Numerical simulations incorporating realistic electrical noise and mechanical vibration noise were conducted to evaluate the algorithm's performance. The results demonstrate that the IPDI_P method provides reliable arrival picks even when SNR values fall below 30 dB, outperforming traditional STA/LTA and AIC approaches.

Field applications using MS data from a deep tunnel project confirmed the effectiveness of the method. The approach successfully identified P-wave arrivals in signals where manual interpretation was difficult, thereby improving the reliability of MS event location.

9.3 S-Wave Arrival Time Identification (Chap. 4)

In addition to P-wave arrival times, accurate S-wave picking is also critical for improving the precision of MS source location. However, S-waves are often difficult to identify in tunnel MS signals because their arrivals may be masked by P-wave energy or background noise.

Chapter 4 extends the phase difference concept introduced in the previous chapter and develops the IPDI_S method. Similar to the P-wave approach, this method analyzes the instantaneous phase relationship between signals recorded by different sensors. Peaks in the IPDI_S function correspond to the onset of S-wave arrivals.

Synthetic signal tests show that the method can successfully identify S-wave arrivals even when they are obscured by P-wave energy. Field applications in deep tunnel monitoring further demonstrate that IPDI_S can detect arrivals that are nearly impossible to identify through visual inspection. By providing more reliable S-wave arrival times, the method significantly improves the accuracy of ms event location and enhances the interpretation of rockburst-related seismic activity.

9.4 MS Event Cluster Location Using Coda Wave Interferometry (Chap. 5)

Precise localization of MS events is essential for identifying active fracture zones and assessing rockburst hazards. Traditional location methods require dense sensor arrays and accurate velocity models, conditions that are often difficult to achieve in underground tunnels.

Chapter 5 introduces an alternative approach based on CWI for locating clusters of MS events. Instead of relying solely on direct wave arrivals, the method utilizes coda waves—the scattered wavefield that follows the primary seismic arrivals. Because coda waves sample multiple propagation paths through the rock mass, they contain rich information about relative event locations.

The proposed method combines three steps:

1. Event clustering based on waveform similarity.
2. Relative location of events using travel-time perturbations extracted from coda wave correlations.
3. Conversion of relative coordinates to absolute positions using reference events.

Numerical simulations indicate that the method can achieve an average location error of approximately 1.46 m. Field tests conducted in deep tunnels show location errors of approximately 3 m for blasting events and slightly higher for rockburst events. Compared with traditional methods, the CWI approach produces tighter event clusters that correspond more closely with observed rock damage zones. This confirms that coda wave information can significantly enhance localization accuracy, particularly in environments with sparse sensor coverage.

9.5 MS Source Location with Updated Heterogeneous Velocity Models (Chap. 6)

Accurate MS event location also depends on reliable knowledge of the seismic velocity structure of the surrounding rock. In deep tunnels, the velocity field changes continuously as excavation progresses and stress redistribution occurs. Conventional methods that assume a homogeneous velocity model can therefore produce large location errors.

Chapter 6 proposes a dynamic velocity inversion approach that incorporates drilling and blasting data to update the velocity model during excavation. Surface wave inversion is first used to estimate shear-wave velocity structures, after which AST is applied to reconstruct a heterogeneous P-wave velocity field.

Once the velocity model is updated, MS events are located using an optimized double-difference algorithm that minimizes travel-time residuals between observed and calculated arrival times. Numerical experiments demonstrate that incorporating heterogeneous velocity structures significantly improves location accuracy.

Field validation was conducted in deep tunnels of the Jinping underground laboratory. Known blasting sources were used as benchmarks to verify the inversion results. The updated velocity models produced more accurate source locations than conventional homogeneous models, confirming the practical effectiveness of the method.

9.6 Stability Analysis of Deep Cross-Tunnels (Chap. 7)

Beyond event detection and localization, MS monitoring data can also provide insight into the mechanical behavior of surrounding rock masses. Chap. 7 integrates velocity inversion results with numerical modeling to evaluate the stability of deep cross tunnels.

The methodology begins with seismic velocity inversion using adjoint-state travel-time tomography. The obtained P-wave velocities are then used to update the elastic modulus of the rock mass, allowing the mechanical properties to evolve dynamically as excavation progresses. These updated parameters are incorporated into numerical simulations using the CASRock program.

Rock failure processes are evaluated using indicators such as equivalent plastic shear strain and local energy release rate. The analysis reveals that failure zones tend to concentrate near tunnel sidewalls and exhibit spatial variations related to the distance between adjacent tunnels. Specifically, a pattern of alternating high- and low-risk zones appears along the tunnel alignment.

The results highlight the importance of integrating geophysical monitoring data with geomechanical analysis to achieve a more comprehensive understanding of rock stability in deep underground construction.

9.7 Dynamic Tomography of Three-Dimensional Velocity Fields (Chap. 8)

Chapter 8 addresses the challenge of reconstructing high-resolution velocity models for large underground caverns. Conventional time-lapse tomography methods often suffer from poor resolution due to unfavorable sensor-source geometries and limited data coverage.

To overcome these limitations, this chapter proposes an SDD tomography method for the dynamic inversion of three-dimensional velocity fields using MS events. The approach combines two key strategies:

1. Double-difference travel-time inversion reduces errors caused by unknown origin times by analyzing relative arrival time differences between event pairs.
2. Stratified inversion divides the velocity model into layers to improve inversion stability and mitigate ill-conditioning caused by uneven sensor distributions.

Resolution tests using synthetic checkerboard models demonstrate that the method can resolve velocity structures with a spatial resolution of approximately 50 meters. Additional synthetic tests simulating excavation processes confirm that the method remains stable even when the dataset contains biased or incomplete observations.

Application of the method to MS monitoring data from the Jinping underground laboratory further demonstrates its capability to reconstruct dynamic velocity changes during excavation. The results show strong agreement between inverted models and the true velocity distribution, confirming the method's robustness for large-scale underground engineering projects.

9.8 Overall Contributions

Together, Chaps. 2–8 form a comprehensive technological framework for MS monitoring in deep tunnel engineering. The framework systematically addresses key challenges associated with noisy monitoring environments, sparse sensor networks, and evolving geological conditions.

The main contributions include the following:

1. An automated method for identifying rock fracture signals without large training datasets.
2. Robust phase picking techniques capable of extracting P- and S-wave arrivals from low-SNR signals.
3. Improved event localization methods using coda wave interferometry and heterogeneous velocity models.
4. Dynamic velocity inversion techniques that adapt to changing excavation conditions.
5. Integrated geophysical-mechanical analysis for evaluating tunnel stability.
6. A novel stratified double-difference tomography method for reconstructing three-dimensional velocity structures.

By integrating these techniques into a unified workflow, the research significantly improves the accuracy and efficiency of MS data processing. The proposed methods provide valuable tools for real-time monitoring of rock fracturing processes and contribute to more reliable rockburst early warning and risk management in deep underground engineering.

Chapter 10
The Core Code of Methods Proposed in This Book

10.1 Declaration

This chapter provides the core codes using MATLAB that are related to the methods proposed in this book.

10.2 Automatic Rock Fracture Signal Identification in Tunnel

```
function [outcome]=Signal_mathematical_model(yy)
% yy is the waveform that needs to be identified
yy_abs=abs(yy);
yy_nor=yy_abs./max(yy_abs);
yy_stan=zeros;
[idx,C]=kmeans(yy_nor,2);
%k-means analysis. We can use the 'kmeans' in MATLAB or 'km' whiten
by the authors
denoise=2*min(C);
if denoise >0.1
  denoise=0.1;
end
for i=1:1:length(yy_nor)
        if yy_nor(i,1)>denoise || yy_nor(i,1)==denoise
         yy_stan(i,1)=yy_nor(i,1);
        end
        if yy_nor(i,1)<denoise
         yy_stan(i,1)=0;
        end
end
%***********First extraction of maximum value***************
```

© Tongji University Press Co., Ltd 2026

Y. Xiao et al., *Near-Field Microseismic Data Processing in Deep Tunnels*,
https://doi.org/10.1007/978-981-92-3208-6_10

```
E_1=zeros;c=1;E_1x=zeros;
for i=2:1:length(yy_stan)-1
        if    yy_stan(i,1)>yy_stan(i-1)    &&    yy_stan(i,1)>yy_
stan(i+1)
                E_1(c,1)=yy_stan(i,1);
                E_1x(c,1)=i;
                c=c+1;
        end
end
%************Second extraction of maximum value***************
E_2x=zeros;E_2y=zeros;c=1; E_2xx=zeros;
for i=2:1:length(E_1)-1
if E_1(i,1)>E_1(i-1) && E_1(i,1)>E_1(i+1)
                E_2y(c,1)=E_1(i,1);
                E_2x(c,1)=E_1x(i,1);
                E_2xx(c,1)=i;
                c=c+1;
   end
end
%************Correction of Maximum Value***************
Modify_y=zeros;x_int=zeros;
x1=E_1x(1,1); x2=E_2x(1,1);
y1=E_1(1,1);y2=E_2y(1,1);
Position_1=find(E_1x==x1);
Position_2=find(E_1x==x2);
Quantity=Position_2-Position_1-1;
Modify_y(1,1)=E_1(1,1);
if Quantity~=0
   x_int(1:Quantity,1)=E_1x(Position_1+1:Position_2-1,1);
   for j=1:1:Quantity
                Modify_y(length(Modify_y)+1,1)=y1+((y2-y1)/
(x2- x1))*(x_int(j,1)- x1);
   end
end
for i=2:1:length(E_2x)
% for i=2:1:3
         x1=E_2x(i-1,1); x2=E_2x(i,1);
        y1=E_2y(i-1,1);y2=E_2y(i,1);
        Position_1=find(E_1x==x1);
        Position_2=find(E_1x==x2);
        Quantity=Position_2-Position_1-1;
        Modify_y(length(Modify_y)+1,1)=E_2y(i-1,1);
                if Quantity~=0
                        x_int=zeros;
                        x_int(1:Quantity,1)=E_1x(Position_
1+1:Position_2-1,1);
                        for j=1:1:Quantity
                                Modify_y(length(Modify_
y)+1,1)=y1+((y2-y1)/(x2- x1))*(x_int(j,1)- x1);
                        end
                end
end
x_f=E_2x(length(E_2x),1);
Position_f=find(E_1x==x_f);
```

```
Modify_y(length(Modify_y)+1,1)=E_1(Position_f,1);
Lack_y=E_1(Position_f+1:end,1);
Modify_y(length(Modify_y)+1:length(Modify_y)+1+length(Lack_
y)-1,1)=Lack_y;
%************Find the value of P***************
y=Modify_y;
big_02=find(Modify_y(:,1)>0.2);
win=zeros;c=1;
for i=1:1:length(big_02)-1
        if i==1
                win(1,1)=big_02(1,1);
        end
        if big_02(i+1,1)-big_02(i,1)==1
                win(c,2)=big_02(i+1,1);
        end
        if big_02(i+1,1)-big_02(i,1)~=1
                win(c+1,1)=big_02(i+1,1);
                win(c+1,2)=big_02(i+1,1);
                c=c+1;
        end
end
winn=zeros;c=1;
for i=1:1:length(win(:,1))
        if win(i,1)<win(i,2)
                if max(Modify_y(win(i,1):win(i,2)))>0.5
                        winn(c,1:2)=win(i,1:2);
                        c=c+1;
                end
        end
end
P=zeros;
for i=1:1:length(winn(:,1))
        po=find(Modify_y==max(Modify_y(winn(i,1):winn(i,2))));
        if po==1
                P(i,1)=po(1,1);
        end
        if po~=1
            poo=find(Modify_y(winn(i,1):winn(i,2))==max(Modify_
y(winn(i,1):winn(i,2))));
                P(i,1)=winn(i,1)+poo-1;
        end
end
Peak_quantity=length(P);
%****************Find the value of F and S*********************
if Peak_quantity==1
        S=find(y(1:winn(1,1))<0.2, 1, 'last');
        F=find(y(winn(1,2):length(y),1)<0.2,
1,'first')+winn(1,2)-1;
        if isempty(F)
                F=find(y==min(y(winn(1,2):length(y))));
        end
        if isempty(S)
         S=find(y==min(y(1:winn(1,1))));
        end
```

```
        gx(1,1)=1;gx(2,1)=E_1x(S,1);gx(3,1)=E_
1x(P,1);gx(4,1)=E_1x(F,1);
        gy(1,1)=0;gy(2,1)=E_1(S,1);gy(3,1)=E_1(P,1);gy(4,1)=E_
1(F,1);
        gx(5,1)=length(yy);gy(5,1)=yy_nor(end,1);
end
if Peak_quantity~=1
        num_y02=length(find(y<0.2));
        if num_y02~=0
                if winn(1,1)~=1
                 S(1,1)=find(y(1:winn(1,1))<0.2, 1, 'last');
                end
                if winn(1,1)==1
                 S(1,1)=1;
                end
                    F(1,1)=find(y(winn(1,2):winn(1+1,1),1)<0.2,
1,'first')+winn(1,2);
          for i=2:1:Peak_quantity-1
                        S(i,1)=find(y(winn(i-
1,2):winn(i,1))<0.2, 1, 'last')+winn(i-1,2);
                  F(i,1)=find(y(winn(i,2):winn(i+1,1),1)<0.2,
1,'first')+winn(i,2);
                end
                S(Peak_quantity,1)=find(y(winn(Peak_quantity-
1,2):winn(Peak_quantity,1))<0.2,1,  'last')+winn(Peak_quantity-
1,2);
                if winn(Peak_quantity,2)==length(y)
                        F(Peak_quantity,1)=length(y);
                end
          if winn(Peak_quantity,2)~=length(y)
                           F(Peak_quantity,1)=find(y(winn(Peak_
quantity,2):length(y),1)<0.2, 1,'first')+winn(Peak_quantity,2)-
1;
          end
          if isempty(F)
                    F(Peak_quantity,1)=find(y==min(y(winn(Peak_
quantity,2):length(y))));
          end
  end
        gxx(1,1)=1;
        gxx(2:1+Peak_quantity,1)=E_1x(F(1:Peak_quantity,1));
        gxx(2+Peak_quantity:1+Peak_quantity*2,1)=E_1x(P(:,1));
        gxx(2+Peak_quantity*2:1+Peak_quantity*3,1)=E_
1x(S(:,1));
        gxx(length(gxx)+1,1)=length(yy);
        gx=sort(gxx);
        gy(1:length(gx))=yy_nor(gx(1:length(gx)));
end
%***********outcome***************
% Outcome the value
end
-----------------------------------------------------------
-----------------------------------------------------------
Function [z, whi]=km(x,n)
```

```
        c1=0.1;c2=0.3;
        whi=zeros(length(x),1);
        l1=abs(x-c1);l2=abs(x-c2);
         x1=zeros; x2=zeros;
        for j=1:1:n %
                %********Cluster***************
                m=1;n=1;
                for i=1:1:length(x)
                        if l1(i,1)<l2(i,1)|| l1(i,1)==l2(i,1)
                                whi(i,1)=1;
                                 x1(m,1)=x(i,1);
                             m=m+1;
                end
                if l1(i,1)>l2(i,1)
                        whi(i,1)=2;
                         x2(n,1)=x(i,1);
                        n=n+1;
                end
        end
        %**************************

      %********Calculate the center of the cluster********

        c1=sum(x1)/length(x1);
        c2=sum(x2)/length(x2);
        l1=abs(x-c1);l2=abs(x-c2);
    end
    z(1,1)=c1;z(2,1)=c2;
end
```

10.3 MS Event Cluster Location in Tunnels Based on Coda Wave Interference

We do not provide the core function of CWI, which can be found in Zhao and Curtis (2019). This section provides the preprocessing of CWI.

```
clear, clc, close all
num_channels_jj=3; %The number of sensors
%%
%**********Extract the arrival times of waveforms and corresponding
sensors*************
arrival_time=zeros; signal=cell(0, 0);
sensors(1:15,1)=31; sensors(1:15,2)=21; sensors(1:15,3)=11;
for jj=1:1:15
        numbera=num2str(jj);
        name=['Source' numbera '.mat'];
        load(name)
```

```
        signal{jj,1}=recordxxx(:,273);
        signal{jj,2}=recordxxx(:,307);
        signal{jj,3}=recordxxx(:,340);
        %calculate the arrival time of P-wave
        for ii=1:1:num_channels_jj
              a_arr1=signal{jj, ii};
              [a_arr1]=bp_filter(a_arr1,0.000025,50,55,600,650);
              aic=AIC(a_arr1);
              P_arr=find(a_arr1==max(a_arr1(1:1000)));
              arrival_time(jj, ii)=P_arr;
        end
end
save('signal_arrival.mat', 'signal', 'arrival_time','sensors');
%%
%****** Select events, and calculate the interrelationships between
waveforms for event clustering*********
clear, clc, close all
sensors_select=zeros; %The sensors selected
sensors_sort=zeros;         % Sorted sensors
signal_sort=cell(0,0); % Matrix of sensors corresponding to sorted
signals
arrival_time_sort=zeros;% Corresponding to the sorted sensor
matrix at that time
event_pair=zeros;          %All signal pairs
max_cr_matrix=zeros; %The cross-correlation of signal pairs
max_cr_average=zeros; % The mean of all events on all channels of
xcorr
max_time_lag=10;
load('signal_arrival.mat');
%********Determine the sensors used for analysis*************
num_event_select=length(sensors(:,1)); %The number of sensors
selected
sensors=sensors(:,1:3);
sensors_name=unique(sensors);
sensors_name=sensors_name(sensors_name~=0);
c=1;
for i=1:1:length(sensors_name)
      if length(find(sensors==sensors_name(i,1)))==length(sensors
(:,1))
                sensors_select(c,1)=sensors_name(i,1);
                c=c+1;
        end
end
num_pairs=(num_event_select-1)*num_event_select/2;
num_sensors_select=length(sensors_select);
for j=1:1:num_sensors_select
        s_name=sensors_select(j,1);
        for i=1:1:num_event_select
                position=find(sensors(i,:)==s_name);
                sensors_sort(i,j)=sensors(i, position);
                signal_sort{i,j}=signal{i, position};
                arrival_time_sort(i,j)=arrival_time(i,
position);
        end
```

```
end
%*******Calculate the xccor of each signal pair*************
c=1;
maxlag=ceil(max_time_lag/(1/6000));
for kk=1:1:num_sensors_select
        d=1;
        for i=1:1:num_event_select-1
                xcorr_start1=arrival_time_sort(i, kk);
                S1=signal_sort{i, kk};
                for j=i+1:1:num_event_select
                        if kk==1
                                event_pair(c,1)=i;
                                event_pair(c,2)=j;
                                c=c+1;
                        end
                        xcorr_start2=arrival_time_sort(j, kk);
                        S2=signal_sort{j, kk};
                        [cr,                lgs]=xcorr(S1(xcorr_
start1:xcorr_start1+1000),S2(xcorr_start2:xcorr_
start2+1000),maxlag,'coeff');
                        max_cr_matrix(d, kk)=max(cr);
                        d=d+1;
                end
        end
end
for i=1:1:num_pairs
        max_cr_average(i,1)=mean(max_cr_matrix(i,:));
end
[max_cr_average_sort, ind]=sort(max_cr_average,'descend');
Max_cr_pair=[event_pair(ind,:)    max_cr_average_sort    max_cr_
matrix(ind,:)];
%*******Save the data*************
needsave.Max_cr_pair=Max_cr_pair;
needsave.path=path;
needsave.Signal=signal_sort;
needsave.arrival_time=arrival_time_sort;
needsave.sensors_sort=sensors_sort;
% needsave.information0714=information0714;
% needsave.information0715=information0715;
needsave.num_sensors_select=num_sensors_select;
save('part2', 'needsave');
%%
% Cluster analysis, which can refer to Zhao and Curtis (2019)
%%
% %***************Estimate the source separation *************
clear, clc, close all
load("part2.mat");
MIN_corr=0.2;
time=(0:0.000025:1)';
dt=0.000025;
veloc=[4500 2800];
source_type='doublecouple';
index=find(needsave.Max_cr_pair(:,3)<0.5, 1, 'first');
needsave.Max_cr_pair(index:end,:)=[];
```

```
pair=needsave.Max_cr_pair(:,1:2);
Signal=needsave.Signal;
num_pairs=length(pair);
num_sensors_select=needsave.num_sensors_select;
win_num=4;          %The number of windows
ideallength=250; %The length of windows
win_start=250;          %The start point after the P-wave arrival
time
search_range=40;
num_interpo_point=20;
sep_pair=zeros;
%******************Calculate      the      source      separa-
tion**************
c=1;
for j=1:1:num_sensors_select
        d=1;
        sep_mean=zeros;
        sep_std=zeros;
        SEP=zeros;
        for i=1:1:13
                ind_event_1=pair(i,1);
                for k=i+1:1:14
                        ind_event_2=pair(k,2);
                        S1=Signal{ind_event_1,1};
                        S2=Signal{ind_event_1,1};
                        arrival_time_1=needsave.arrival_
time(ind_event_1,1)
                        arrival_time_2=needsave.arrival_
time(ind_event_2,1)
                        S1=S1(arrival_time_1:end,1);
                        S2=S2(arrival_time_2:end,1);
                        %************Determine     the     win_
length*************
                        residual_length_1=length(S1(:,1))-win_
start;
                        residual_length_2=length(S2(:,1))-win_
start;
                        win_length_total=min(residual_length_
1,residual_length_2);
                        if win_length_total>win_num*ideallength
                                win_length_total=win_
num*ideallength;
                        end
                        win_length_total_s=win_length_
total*dt;
                        win_start_s=win_start*dt;
                        win_end=win_start+win_length_total;
                        win_end_s=win_end*dt;
                        each_win_length_s=win_length_total_s/
win_num;
                        [SEP(d,1:win_num),  sep_mean(d),  sep_
std(d)]=CWI_sep(S1,S2,time, dt, veloc, source_type, …
                        each_win_length_s, win_start_s, win_end_
s, search_range, num_interpo_point);
```

```
                    d=d+1;
                if j==1
                        sep_pair(c,1)=i;
                        sep_pair(c,2)=k;
                        c=c+1;
                    end
                end
        end
        sensors_name=needsave.sensors_sort(1,j);
        eval(['SEP_' num2str(sensors_name) '=SEP;']);
        eval(['sep_mean' num2str(sensors_name) '=sep_mean;']);
        eval(['sep_std' num2str(sensors_name) '=sep_std;']);
end
save('part3.mat');
%%
% %****************The data for source location*************
clear, clc, close all
Frequency=zeros;
load('part3.mat')
num_sensors_select=needsave.num_sensors_select;
num_event_select=length(needsave.path);
fre=zeros;
for i=1:1:num_sensors_select
        for j=1:1:15
                si=Signal{j,i};
                [Y, f]=FFTAnalysis(si, 1/6000);
                [maxValue, maxIndex]=max(Y);
                fre(j,i)=f(maxIndex);
        end
        free=fre(:,i);
        maxfre_index=find(free==max(free));
        free(maxfre_index)=[];
        minfre_index=find(free==min(free));
        free(minfre_index)=[];
        Frequency(i,1)=mean(free(:,1));
end
domin_wavelength=veloc(2)./Frequency;
nevent=num_event_select; % number of events to locate
num_channel=num_sensors_select;% number of channels to use
num_pair=length(sep_pair); % compute number of event pairs
max_iteration=240;% maximum number of iterations allowed
L_tolerance=1e-5;%minimum decrease in the objective function,
below which the iteration shall be terminated
% Separation data
seps11=[sep_mean11' sep_std11'];
seps21=[sep_mean21' sep_std21'];
seps31=[sep_mean31' sep_std31'];% load separation data for loca-
tion
Mu_n=[seps11(:,1) seps21(:,1) seps31(:,1)];% means of separation
estimated
SIGMA_n=[seps11(:,2) seps21(:,2) seps31(:,2)]; % standard devia-
tions of separation estimated
Appli_up_bound_mu=0.60;          Appli_low_bound_mu=0;
Appli_up_bound_sigma=0.17; Appli_low_bound_sigma=0.001;
```

```
% Prepare data
sep_mean=Mu_n;
sep_std=SIGMA_n;
sep_nor=zeros(num_pair, num_channel);
sep_nor_std=zeros(num_pair, num_channel);
for ii=1:num_channel
        sep_nor(:,ii)=sep_mean(:,ii)/domin_wavelength(ii);
        sep_nor_std(:,ii)=sep_std(:,ii)/domin_wavelength(ii);
end
% Identify unsuitable event pairs
available_sep_index_1=ones(num_channel,1);
for jj=1:num_channel
        ii=1;
for ii=1:num_pair
        if  sep_nor(ii,  jj)<=Appli_up_bound_mu  &  sep_nor(ii,
jj)>Appli_low_bound_mu…
                                sep_nor_std(ii,    jj)<=Appli_up_bound_
sigma & sep_nor_std(ii, jj)>Appli_up_bound_sigma;
                sep_nor_Reduced(available_sep_index_1(jj),
jj)=sep_nor(ii, jj);
                sep_nor_std_Reduced(available_sep_index_1(jj),
jj)=sep_nor_std(ii, jj);
                available_sep_index_1(jj)=available_sep_index_
1(jj)+1;
        else
        end
end
        available_sep_index_1(jj)=available_sep_index_1(jj)-1;
end
% Clarify the indexes of the discarded pairs to be fed into function
L
num_discarded_pairs=zeros(1,num_channel);
for jj=1:num_channel
        num_discarded_pairs(jj)=numel(find(sep_
nor(:,jj)>Appli_up_bound_mu   |   sep_nor(:,jj)<Appli_low_bound_
mu | …
        sep_nor_std(:,jj)>Appli_up_bound_sigma     |     sep_nor_
std(:,jj)<Appli_low_bound_sigma));
        jj=jj+1;
end
discarded_pairs=zeros(max(num_discarded_pairs), num_channel);
for jj=1:num_channel
        if num_discarded_pairs(jj)~=0
                discarded_pairs(1:num_discarded_pairs(jj),
jj)=find(sep_nor(:,jj)>Appli_up_bound_mu | sep_nor(:,jj)<Appli_
low_bound_mu | …
              sep_nor_std(:,jj)>Appli_up_bound_sigma | sep_nor_
std(:,jj)<Appli_low_bound_sigma);
        else
        end
end
 save('part4.mat');
%%
% ******************Source location*************
```

```
clear, clc, close all
load('part4.mat')
for_location.needsave=needsave;
for_location.Mu_n=Mu_n;
for_location.SIGMA_n=SIGMA_n;
for_location.domin_wavelength=domin_wavelength;
for_location.discarded_pairs=discarded_pairs;
for_location.max_iteration=max_iteration;
for_location.L_tolerance=L_tolerance;
for_location.nevent=nevent;
clearvars -except for_location
MU_n=for_location. Mu_n;
nevent=for_location.nevent;
SIGMA_n=for_location. SIGMA_n;
domin_wavelength=for_location.domin_wavelength;
discarded_pairs=for_location.discarded_pairs;
max_iteration=for_location.max_iteration;
L_tolerance=for_location. L_tolerance;
initial_position=[px;py;ones(1,15)]';
[   optm_loc,   Func_L_add]=Source_Location(initial_position,MU_
n,SIGMA_n,…
  domin_wavelength, discarded_pairs, max_iteration,L_tolerance);
```

The Akaike information criterion (AIC) is an automatic method to pick P-wave arrival times. The core code of this method is shown below:

```
function aic=AIC(x)
% 'x' is the signal
   aic=zeros;
   for i=1:1:length(x)-1
                    aic(i,1)=i*log10(var(x(1:i,1)))+(length(x)-i-
1)*log10(var(x(i+1:length(x),1)));
   end
end
```

The above code can calculate the relative location of the MS event cluster. The core code that converts the relative location to absolute location is shown below:

```
clear, clc, close all
%-----------------Parameter--------------------
P_final=[]; % The relative location of reference events
P_ini=[]; The absolute location of reference events
number_of_points=3; % The number of points need to be converted
individual=1;
translation_range=[0,50];
```

```
rotation_range=[0,360];
Mutation_rate=1;
Number_of_iteration=10;
%----------------GA------------------
center=[mean(P_final(:,1)),     mean(P_final(:,2)),     mean(P_
final(:,3))];
gene_old(1:3,1:individual)=randi([0, 50],3,individual);
gene_old(4:6,1:individual)=randi([0, 360],3,individual);
[P_new]==ro_and_train(gene_old(4,1),gene_old(5,1),gene_
old(6,1),gene_old(1,1),gene_old(2,1),gene_old(3,1),P_final);
fitness(1,1)=(P_final(1,1)-P_new(1,1))^2+(P_final(1,2)-P_
new(1,2))^2+(P_final(1,3)-P_new(1,3))^2+...
          (P_final(2,1)-P_new(2,1))^2+(P_final(2,2)-P_
new(2,2))^2+(P_final(2,3)-P_new(2,3))^2+...
          (P_final(3,1)-P_new(3,1))^2+(P_final(3,2)-P_
new(3,2))^2+(P_final(3,3)-P_new(3,3))^2;
gene_new=zeros;
for i=2:1:Number_of_iteration
          % Mutation
          Mutation;
          [P_new]=ro_and_train(gene_new(4,1),gene_new(5,1),gene_
new(6,1),gene_new(1,1),gene_new(2,1),gene_new(3,1),P_final);
          % calculate the fitness of old and new
          ind_fitness=(P_final(1,1)-P_new(1,1))^2+(P_final(1,2)-
P_new(1,2))^2+(P_final(1,3)-P_new(1,3))^2+...
          (P_final(2,1)-P_new(2,1))^2+(P_final(2,2)-P_
new(2,2))^2+(P_final(2,3)-P_new(2,3))^2+...
          (P_final(3,1)-P_new(3,1))^2+(P_final(3,2)-P_
new(3,2))^2+(P_final(3,3)-P_new(3,3))^2;
          if ind_fitness>fitness(i-1,1)
                    gene_old=gene_old;
                    fitness(i,1)=fitness(i-1,1);
          else
                    gene_old=gene_new;
                    fitness(i,1)=ind_fitness;
          end
end
```

The code of 'Mutation' is shown below:

```
j=1;
r=randi([0, 1]);
if r>0.6666
          gene_new(1,j)=gene_old(1,j)+Mutation_rate;
elseif r<0.3333
          gene_new(1,j)=gene_old(1,j)-Mutation_rate;
elseif r>0.3333 && r<0.6666
          gene_new(1,j)=gene_old(1,j);
end
r=randi([0, 1]);
if r>0.6666
          gene_new(2,j)=gene_old(2,j)+Mutation_rate;
```

```
elseif r<0.3333
        gene_new(2,j)=gene_old(2,j)-Mutation_rate;
elseif r>0.3333 && r<0.6666
        gene_new(2,j)=gene_old(2,j);
end
r=randi([0, 1]);
if r>0.6666
        gene_new(3,j)=gene_old(3,j)+Mutation_rate;
elseif r<0.3333
        gene_new(3,j)=gene_old(3,j)-Mutation_rate;
elseif r>0.3333 && r<0.6666
        gene_new(3,j)=gene_old(3,j);
end
r=randi([0, 1]);
if r>0.6666
        gene_new(4,j)=gene_old(4,j)+Mutation_rate;
elseif r<0.3333
        gene_new(4,j)=gene_old(4,j)-Mutation_rate;
elseif r>0.3333 && r<0.6666
        gene_new(4,j)=gene_old(4,j);
end
r=randi([0, 1]);
if r>0.6666
        gene_new(5,j)=gene_old(5,j)+Mutation_rate;
elseif r<0.3333
        gene_new(5,j)=gene_old(5,j)-Mutation_rate;
elseif r>0.3333 && r<0.6666
        gene_new(5,j)=gene_old(5,j);
end
r=randi([0, 1]);
if r>0.6666
        gene_new(6,j)=gene_old(6,j)+Mutation_rate;
elseif r<0.3333
        gene_new(6,j)=gene_old(6,j)-Mutation_rate;
elseif r>0.3333 && r<0.6666
        gene_new(6,j)=gene_old(6,j);
end
```

The function of 'ro_and_train' that rotates and translates the location is shown below:

```
function [P_new]=ro_and_train(rx, ry, rz, tx, ty, tz,P_final)
p1=P_final(1,:);
p2=P_final(2,:);
p3=P_final(3,:);
P=[p1; p2; p3];
center=mean(P,1);          % (xa, ya, za)
P_shift=P - center;
rx=deg2rad(rx);
ry=deg2rad(ry);
rz=deg2rad(rz);
Rx=[1 0 0;
```

```
        0 cos(rx) -sin(rx);
        0 sin(rx) cos(rx)];
Ry=[cos(ry) 0 sin(ry);
        0 1 0;
        -sin(ry) 0 cos(ry)];
Rz=[cos(rz) -sin(rz) 0;
        sin(rz) cos(rz) 0;
        0 0 1];
R=Rz * Ry * Rx;
P_rot=(R * P_shift')';
P_rot=P_rot+center;
T=[tx, ty, tz];
P_final=P_rot+T;
p1_new=P_final(1,:);
p2_new=P_final(2,:);
p3_new=P_final(3,:);
P_new=[p1_new;p2_new;p3_new];
end
```

10.4 Adjoint-State Travel Time Tomography

The main program

```
clear, clc, close all
%% --------------Parameters--------------
INVMAT=1; % 1 FATT
NX=200; %The number of grid
NY=200;
DH=0.1;
% Time Stepping
TIME=0.008;% time_of_wave_propagation_(in_sec)_(TIME)=6.0
% receivers
recpos(1,1:20)=200;
recpos(2,1:20)=(5:10:195);
recpos(3,1:20)=0;
% source positions
nshots=20; % The number of shots
srcpos(1,1:20)=200;
srcpos(2,1:20)=(5:10:195);
srcpos(3,1:20)=0; % tshift
srcpos(4,1:20)=0;
% Convergence criterion for Eikonal solver
TTNORM=1e-6; %Convergence_criterion_for_fast_sweeping_technique
% RAJZEL inversion parameters
% ITERMAX=600;
ITERMAX=100;
EPS_ADJ=1e-9; % water_level_sweep_adj
SWS_TAPER_GRAD_VERT=0; % apply_vertical_taper_(yes=1)
```

```
SWS_TAPER_GRAD_HOR=0; % apply_horizontal_taper_(yes=1)
EXP_TAPER_GRAD_HOR=0;      %      exponent_of_depth_scaling_for_
preconditioning
SWS_TAPER_CIRCULAR_PER_SHOT=0;     %apply_cylindrical_taper_per_
shot_(yes=1)
VPUPPERLIM=6500; % upper_limit_for_vp/lambda
VPLOWERLIM=3500; % lower_limit_for_vp/lambda
% --- Optimization-Method ---
GRAD_METHOD=2; % (PCG=1/LBFGS=2/GRADIENT=3)
NLBFGS=20; % save_(NLBFGS)_updates_during_LBFGS_optimization
% Step length estimation
LINESEARCH=2; % step_length_estimation_(Wolfe=1/Parabolic=2)
SCALEFAC=2.0; % scale factor for the step length
STEPMAX=10; % maximum_number_of_attemps_to_find_a_step_length
EPS_SCALE=0.01; % maximum_model_change_of_maximum_model_value
% fatt workflow
PRO=0;
% Minimum number of iterations per frequency
MIN_ITER=0;
%% --------------Prepare--------------
sim_data=load(sim_data_dir);
ntr=length(recpos(1,:));
Tmod=zeros(ntr,1);
if INVMAT>0
        Tobs=zeros(ntr,1);
        Tres=zeros(ntr,1);
end
TT=zeros(NY,NX);Vp=zeros(NY,NX);S=zeros(NY,NX);
INFO=0; % If INVMAT!=0 deactivate unnecessary output
if INVMAT==0%If INVMAT==0 allow unnecessary output
        INFO=1;
end
srcpos(1:2,1:nshots)=srcpos(1:2,1:nshots).*DH;
% Vp=sim_data.v;
Vp=ones(200,200).*5000; S=1./Vp;
picktimes=sim_data.travel_sensor;
if INVMAT==1
  fatt; % FATT
end
```

The 'FATT' function is shown below:

```
if GRAD_METHOD==2
        NLBFGS_class=1;                  % number of parameter
classes
        NLBFGS_vec=NLBFGS_class*NX*NY; % length of one LBFGS-
parameter class
        LBFGS_pointer=1;        %initiate pointer in the cyclic
LBFGS-vectors
        y_LBFGS=zeros(NLBFGS_vec*NLBFGS,1);                    s_
LBFGS=zeros(NLBFGS_vec*NLBFGS,1);
        q_LBFGS=zeros(NLBFGS_vec,1);
```

```
        r_LBFGS=zeros(NLBFGS_vec,1);
        rho_LBFGS=zeros(NLBFGS,1);
        alpha_LBFGS=zeros(NLBFGS,1);
        beta_LBFGS=zeros(NLBFGS,1);
end
% memory of L2 norm
L2t=zeros(4,1);
% memory for gradient
lam=zeros(NY, NX);
grad=zeros(NY, NX);
gradm=zeros(NY, NX);
Hgrad=zeros(NY, NX);
Vpnp1=zeros(NY, NX);
iter_true=1;
%% Begin of FATT inversion workflow
iter=1;
while iter<=ITERMAX
        if GRAD_METHOD==2
                % increase pointer to LBFGS-vector
                if iter>2
                        LBFGS_pointer=LBFGS_pointer+1;
                end
                % if LBFGS-pointer>NLBFGS ->set LBFGS_pointer=1
                if LBFGS_pointer>NLBFGS
                        LBFGS_pointer=1;
                end
                fprintf('进度: %d/%d (%0.1f%%)', iter, ITERMAX,
100*iter/ITERMAX);
        end
        % calculate Vp gradient and objective function
        % TT=zeros;
        % Tmod=zeros;
        % Tobs=zeros;
        [L2,grad,S,TT,lam,Tmod,Tobs,Tres, srcpos, recpos]=grad_
obj(grad,S,TT,lam,Tmod,Tobs,Tres, srcpos, nshots, recpos, ntr,
iter,NX,NY,DH,TIME,TTNORM,picktimes,EPS_ADJ);
        L2t(1)=L2;
        % calculate descent directon gradm from gradient grad
        gradm=-1. *grad;
        if GRAD_METHOD==2
                Lbfgs;
        end
        % check if search direction is a descent direction, other-
wise reset l-BFGS history
        check_descent;
        % Estimate optimum step length
        % by inexact parabolic line search
        if LINESEARCH==2
            [eps_scale,Hgrad, grad,Vp,S,TT,lam,Tmod,Tobs,Tres]=
parabolicls(Hgrad,grad,Vp,S,TT,lam,Tmod,Tobs,Tres,srcpos,nshots,
recpos,ntr,iter,L2,NY,NX,MIN_ITER,STEPMAX,SCALEFAC,EPS_
SCALE,VPLOWERLIM,VPUPPERLIM,DH,TIME,TTNORM,picktimes,EPS_ADJ);
        end
        % saving history of final L2
```

```
        L2_hist(iter)=L2t(1);
[Hgrad,Vp,Vpnp1]=calc_mat_change_wolfe(Hgrad,Vp,Vpnp1,alpha,
itest,NX,NY,VPLOWERLIM,VPUPPERLIM);
        S=1./Vp;
        %%
        % calculating difference of the actual L2 and before two
iterations,
        % dividing with L2_hist[iter-2] provide changing in percent
        if iter>2
                diff=abs(L2_hist(iter-2)-L2_hist(iter))/L2_
hist(iter-2);
        end
        % are convergence criteria satisfied
        if diff<=PRO || eps_scale<1e-8
                save(['data\out',num2str(1),'.mat'], 'Vp');
                iter=0;
        end
        iter=iter+1;
        iter_true=iter_true+1;
end
```

'calc_mat_change_wolfe.m' is shown below:

```
function [Hgrad,Vp,Vpnp1]=calc_mat_change_wolfe(Hgrad,Vp,Vpnp1,
alpha,itest,NX,NY,VPLOWERLIM,VPUPPERLIM)
for i=1:1:NX
        for j=1:1:NY
                % P-wave velocity
                Vpnp1(j,i)=Vp(j,i)+alpha * Hgrad(j,i);
                % apply hard constraints
                if Vpnp1(j,i)<VPLOWERLIM
                        Vpnp1(j,i)=Vp(j,i);
                end
                if Vpnp1(j,i)>VPUPPERLIM
                        Vpnp1(j,i)=Vp(j,i);
                end
                % P-wave velocity should not be smaller than zero
                if Vpnp1(j,i)<0.0
                        Vpnp1(j,i)=Vp(j,i);
                end
                if itest==0
                        Vp(j,i)=Vpnp1(j,i);
                end
        end
end
end
```

'calc_opt_setp.m' is shown below:

```
Function [L2t,epst, opteps]=calc_opt_step(L2t,epst, sws)
```

```
critmult=5.0;
n=3; % number of test calculations
A=zeros(n,n);
x=zeros(n,1);
b=zeros(n,1);
% define coefficient matrix A
for i=1:n
        A(i,3)=epst(i)^2;
        A(i,2)=epst(i);
        A(i,1)=1.0;
end
% define RHS vector b
b(:,1)=L2t(1:n);
% LU_decomp(A,x,b,n)
[A,b,x]=solvelin(A,b,x,n,1);
% calculate optimal step length ->extremum of the parabola
opteps=-1*x(2)/(2.0*x(3));
if (2.0*x(3)<0.0) && (L2t(2)>L2t(3))
        opteps=epst(3);
end
if opteps>epst(3)
        opteps=epst(3);
end
if opteps<0.0
        opteps=epst(1);
end
end
```

The 'check_descent.m' is shown below:

```
% check if search direction is a descent direction
nom=sum(Hgrad(:). * grad(:));
denom=sum(grad(:). * grad(:));
p=-nom/denom;
if p<0
        if GRAD_METHOD==2
                for i=1:1:NLBFGS_vec*NLBFGS
                        y_LBFGS(i)=0.0;
                        s_LBFGS(i)=0.0;
                end
        end
end
```

'cp_grad_frame.m' is shown below:

```
function [A]=cp_grad_frame(A,NX,NY)
        % Top
        A(2,:)=A(3,:);
        A(1,:)=A(2,:);
        % Bottom
        A(NY-1,:)=A(NY-2,:);
```

```
        A(NY,:)=A(NY-1,:);
        % Left
        A(:,2)=A(:,3);
        A(:,1)=A(:,2);
        % Right
        A(:,NX-1)=A(:,NX-2);
        A(:,NX)=A(:,NX-1);
end
```

The 'dotp.m' is shown below:

```
function [tmp]=dotp(vec1, vec2, n1, n2, sw)
tmp=0;
if sw==0
        % dot product for vectors with same length/same indexing
        for i=n1:n2
                tmp=tmp+vec1(i) * vec2(i);
        end
end
if sw==1
        % dot product where vec2 starts from 1 while vec1 uses n1:n2
        h=1;
        for i=n1:n2
                tmp=tmp+vec1(i) * vec2(h);
                h=h+1;
        end
end
end
```

'grad_obj.m' is shown below:

```
function [L2,grad,S,TT,lam,Tmod,Tobs,Tres, srcpos, recpos]=grad_
obj(grad,S,TT,lam,Tmod,Tobs,Tres, srcpos, nshots, recpos, ntr,
iter,NX,NY,DH,TIME,TTNORM,picktimes,EPS_ADJ)
grad_shot=zeros(NY,NX);
recflag=zeros(NY,NX);
% suppress modeled FA traveltime output during FATT
SEISMO=0;
% set gradient matrix to zero before next iteration
grad=zeros(NY,NX);
L2=0;
% define receiver flag field
for j=1:1:NY
        for i=1:1:NX
                recflag(j,i)=0;
                for l=1:1:ntr
                        if i==recpos(1,l) && j==recpos(2,l)
                                recflag(j,i)=1;
                        end
                end
        end
end
for ishot=1:1:nshots
```

```
        % for ishot=1:1:1
        % set gradient for each shot to zero before next iteration
        grad_shot=zeros(NY,NX);
        %% solve forward problem with Eikonal solver
        % Eikonal solver
        nxsrc=round(srcpos(1, ishot)/DH);
        nysrc=round(srcpos(2, ishot)/DH);
        TT=ones(NY,NX). *2*TIME;
        % initialize traveltime field at source positions
        TT(nysrc, nxsrc)=0.0;
        % initialize l1 norm of TT - TTold
        lnorm1=10 * TIME;
        % apply fast sweeping method to solve the eikonal equation
        while lnorm1>=TTNORM
          % save old TT values
          TTold=TT;
          % sweep with order according to Zhao (2004)
          [S,TT]=sweep(S,TT,nxsrc, nysrc,1,NX,1,1,NY,1,NY,NX,DH);
          [S,TT]=sweep(S,TT,nxsrc,          nysrc,NX,1,             -
1,1,NY,1,NY,NX,DH);
                [S,TT]=sweep(S,TT,nxsrc, nysrc,NX,1, -1,NY,1, -
1,NY,NX,DH);
                [S,TT]=sweep(S,TT,nxsrc,      nysrc,1,NX,1,NY,1,
-1,NY,NX,DH);
                % calculate l1 norm of TT - TTold
                lnorm1=sum(abs(TT(:) - TTold(:)))/(NX*NY);
        end
        % calculate TT at receiver positions
        for i=1:1:ntr
                Tmod(i,1)=TT(recpos(2,i), recpos(1,i));
        end
        TTold=zeros(NY,NX);
        %% calculate traveltime residuals at receiver positions
        for i=1:1:ntr
                Tobs(i,1)=picktimes(i, ishot);
                if Tobs(i,1)>=0
                        % calculate L2 traveltime residuals
                        Tres(i,1)=Tmod(i,1)- Tobs(i,1);
                        % calculate L2 objective function
                        L2=L2+Tres(i,1) * Tres(i,1);
                        iy=recpos(2,i);
                        ix=recpos(1,i);
                        if iy<=NY-1
                                gradT=(TT(iy+1,   ix)   -   TT(iy,
ix))/DH;
                                if abs(gradT)>0.0
                                        Tres(i,1)=Tres(i,1)/
gradT;
                                else
                                        Tres(i,1)=0.0;
                                end
                        elseif iy==NY
                                gradT=(TT(iy,   ix)   -   TT(iy-1,
ix))/DH;
```

```
                                    if abs(gradT)>0.0
                                            Tres(i,1)=Tres(i,1)/
gradT;
                                    else
                                            Tres(i,1)=0.0;
                                    end
                            end
                    end
        end
        %% calculate adjoint Eikonal wavefield
        % initialize boundary condition for lambda
        eps=0.0;
        lam=zeros(NY,NX);
        for j=2:1:NY-1
                for i=2:1:NX-1
                        lam(j,i)=10*TIME;
                        for l=1:1:ntr
                                if      i==recpos(1,l)       &&
j==recpos(2,l)
                                        lam(j,i)=Tres(l);
                                end
                        end
                end
        end
        % initialize l1 norm of lam - lamold
        lnorm1=10.0 * TIME;
        % apply fast sweeping method to solve the eikonal equation
        while lnorm1>=TTNORM
                lamold=lam;
                % sweep with order according to Taillandier (2009)
                [lam,TT]=sweep_adj(lam,TT,Tres,         recpos,
ntr,2,NX-1,1,2,NY-1,1,recflag,NX,NY,DH,EPS_ADJ);
                [lam,TT]=sweep_adj(lam,TT,Tres,         recpos,
ntr,2,NX-1,1,NY-1,2, -1,recflag,NX,NY,DH,EPS_ADJ);
                [lam,TT]=sweep_adj(lam,TT,Tres, recpos, ntr,NX-
1,2, -1,2,NY-1,1,recflag,NX,NY,DH,EPS_ADJ);
                [lam,TT]=sweep_adj(lam,TT,Tres, recpos, ntr,NX-
1,2, -1,NY-1,2, -1,recflag,NX,NY,DH,EPS_ADJ);
                % calculate l1 norm of lam - lamold
                lnorm1=sum(abs(lam(:) - lamold(:)))/(NX*NY);
                fprintf('lnorm1: %d\n', lnorm1);
        end
        lamold=zeros(NY,NX);
        %% assemble gradient for each shot
        % assemble gradient from one shots
        grad_shot(2:NY-1, 2:NX-1)=lam(2:NY-1, 2:NX-1). * (S(2:NY-
1, 2:NX-1).^3);
        grad_shot=cp_grad_frame(grad_shot,NX,NY);
        grad=grad+grad_shot;
end
grad_shot=zeros(NY,NX);
recflag=zeros(NY,NX);
end
```

The 'LBFGS.m' is shown below:

```
Itershift=1;
if iter>1
        h=NLBFGS_vec*(LBFGS_pointer1)+1;
        jac=load('data\s_p_old.mat');
        jac1=load('data\s_p_vp_old.mat');
        for i=1:1:NX
                for j=1:1:NY
                        % calculate and save y, s at iteration step
iter
                        y_LBFGS(h)=grad(j,i)-jac.grad(j,i);
                        s_LBFGS(h)=Vp(j,i)-jac1. Vp(j,i);
                        h=h+1;
                end
        end
        % calculate improved first guess Hessian gamma_LBFGS
        h1=NLBFGS_vec*(LBFGS_pointer1)+1;
        h2=NLBFGS_vec*LBFGS_pointer;
        sum_nom=dotp(y_LBFGS, s_LBFGS,h1,h2,0);
        sum_denom=dotp(y_LBFGS,y_LBFGS,h1,h2,0);
        gamma_LBFGS=sum_nom/sum_denom;
        % update variable rho for all LBFGS-iterations and all
parameter classes
        for k=1:1:NLBFGS
                h1=NLBFGS_vec*(k-1)+1;
                h2=NLBFGS_vec*k;
                sum_nom=dotp(y_LBFGS, s_LBFGS,h1,h2,0);
                if abs(sum_nom)>0.0
                        rho_LBFGS(k)=1.0/sum_nom;
                else
                        rho_LBFGS(k)=0.0;
                end
        end
        % save q_LBFGS for all material parameters
        h=1;
        for i=1:1:NX
                for j=1:1:NY
                        q_LBFGS(h)=gradm(j,i);
                        h=h+1;
                end
        end
        % update alpha_LBFGS and q_LBFGS
        for k=NLBFGS:-1:1
                h1=NLBFGS_vec*(k-1)+1;
                h2=NLBFGS_vec*k;
                sum_nom=dotp(s_LBFGS,q_LBFGS,h1,h2,1);
                alpha_LBFGS(k)=rho_LBFGS(k) * sum_nom;
                % update q for all material parameters
                h=NLBFGS_vec*(k-1)+1;
                for i=1:1:NLBFGS
```

```
                        q_LBFGS(i)=q_LBFGS(i) - alpha_LBFGS(k) *
y_LBFGS(h);
                        h=h+1;
                end
        end
        % Multiply gradient with approximated Hessian
        for i=1:1:NLBFGS
                r_LBFGS(i)=gamma_LBFGS * q_LBFGS(i);
        end
        % calculate H^-1 * gradm[j][i]
        for k=1:NLBFGS
                h1=NLBFGS_vec*(k-1)+1;
                h2=NLBFGS_vec*k;
                % calculate beta_LBFGS
                sum_nom=dotp(y_LBFGS,r_LBFGS,h1,h2,1);
                beta_LBFGS_1       =rho_LBFGS(k) * sum_nom;
                h=h1;
                for i=1:NLBFGS_vec
                        r_LBFGS(i)=r_LBFGS(i)+s_LBFGS(h)        *
(alpha_LBFGS(k) - beta_LBFGS_1);
                        h=h+1;
                end
        end
        % update gradients
        h=1;
        % Vp
        for i=1:1:NX
                for j=1:1:NY
                        Hgrad(j,i)=r_LBFGS(h);
                        h=h+1;
                end
        end
        if iter==1
                Hgrad=gradm;
        end
        %%
        % save old models Vp
        save('s_p_vp_old.mat', 'Vp');
        % save old gradient
        save('s_p_old.mat', 'grad');
        % save H^-1 * g
        save('s_c_old.mat', 'Hgrad');
end
```

The 'parabolicls' are shown below:

```
function    [eps_scale,Hgrad,grad,Vp,S,TT,lam,Tmod,Tobs,Tres]=
parabolicls(Hgrad,grad,Vp,S,TT,lam,Tmod,Tobs,Tres,srcpos,nshots,
recpos,ntr,iter,L2,NY,NX,MIN_ITER,STEPMAX,SCALEFAC,EPS_
SCALE,VPLOWERLIM,VPUPPERLIM,DH,TIME,TTNORM,picktimes,EPS_ADJ)
Vpnp1=zeros(NY,NX);
Snp1=zeros(NY,NX);
```

```
gt=zeros(NY,NX);
L2t=zeros(3,1);
epst1=zeros(3,1);
scalefac=SCALEFAC;
stepmax=STEPMAX; % number of maximum misfit calculations/
steplength 2/3
step1=0;
step2=0;
% start with first guess for step length alpha
maxgrad=max(Hgrad(:));
maxvp=max(Vp(:));
alpha=EPS_SCALE * abs(maxvp/maxgrad);
countstep=0; % count number of forward calculations
L2t(1)=L2;
itests=2;
iteste=2;
while step1~=1 || step2~=1
        for itest=2:1:iteste
                % test Vp-update
[Hgrad,Vp,Vpnp1]=calc_mat_change_wolfe(Hgrad,Vp,Vpnp1,alpha,1,
NX,NY,VPLOWERLIM,VPUPPERLIM);
                % calculate slowness S from test Vp model
                Snp1=1./Vpnp1;
[L2t(itest),    gt,Snp1,TT,lam,Tmod,Tobs,Tres,    srcpos,
recpos]=grad_obj(gt,Snp1,TT,lam,Tmod,Tobs,Tres, srcpos, nshots,
recpos, ntr, iter,NX,NY,DH,TIME,TTNORM,picktimes,EPS_ADJ);
        end
        % Did not found a step size which reduces the misfit function
        if step1==0 && L2t(1)<L2t(2)
                alpha=alpha/scalefac;
                countstep=countstep+1;
        end
        % Found a step size with L2t[2]<L2t[3]
        if step1==1 && L2t(2)<L2t(3)
                epst1(3)=alpha;
                step2=1;
        end
        % Could not found a step size with L2t[2]<L2t[3]*/
        if step1==1 && L2t(2)>=L2t(3)
                epst1(3)=alpha;
                % increase step length to find a larger misfit func-
tion than L2t[2]
                alpha=alpha+(alpha/scalefac);
                countstep=countstep+1;
        end
        % found a step size which reduces the misfit function
        if step1==0 && L2t(1)>L2t(2)
                epst1(2)=alpha;
                step1=1;
                iteste=3;
                itests=3;
                countstep=0;
                % find a second step length with a larger misfit
function than L2t[2]*/
```

```
                alpha=alpha+(alpha/scalefac);
        end
        step3=0;
        if step1==0 && countstep>stepmax
                step3=1;
                break
        end
        if step1==1 && countstep>stepmax
                step1=1;
                step2=1;
        end
end
if step1==1
        [L2t,epst, opteps_vp]=calc_opt_step(L2t,epst1,1);
        eps_scale=opteps_vp;
end
end
```

The 'solvelin.m' is shown below:

```
function [AA,bb,x]=solvelin(AA,bb,x,e, method)
rows=e;
columns=e;
A=zeros (rows, columns);
b=zeros(rows,1);
% temporary variables
A(1:rows, 1:columns)=AA(1:rows, 1:columns);
b(1:rows,1)=bb(1:rows);
if method==1
        for k=1:rows-1
                for n=k:rows-1
                        a=A(n+1, k)/A(k, k);
                        for m=1:columns
                                A(n+1, m)=A(n+1, m) - a * A(k, m);
                        end
                        b(n+1)=b(n+1) - a * b(k);
                end
        end
        for k=rows:-1:1
                c=b(k);
                for m=columns:-1:(k+1)
                        c=c - A(k, m) * x(m);
                end
                x(k)=c/A(k, k);
        end
end
end
```

'sweep_adj.m' is shown below:

```
function    [lam,TT]=sweep_adj(lam,TT,Tres,    recpos,    ntr,
nx1,nx2,ndx, ny1,ny2,ndy, recflag,NX,NY,DH,EPS_ADJ)
```

```
h=nx1;
for i=2:(NX-1)
        k=ny1;
        for j=2:(NY-1)
                if recflag(k,h)==0
                        % assemble Eq. (3.6) in Leung & Qian
(2006).
                        ap=-(TT(k, h+1) - TT(k, h)) /DH;
                        am=-(TT(k, h)       TT(k, h-1))/DH;
                        bp=-(TT(k+1, h) - TT(k, h)) /DH;
                        bm=-(TT(k, h)       TT(k-1, h))/DH;
                        app=(ap+abs(ap))/2.0;
                        apm=(ap - abs(ap))/2.0;
                        amp=(am+abs(am))/2.0;
                        amm=(am - abs(am))/2.0;
                        bpp=(bp+abs(bp))/2.0;
                        bpm=(bp - abs(bp))/2.0;
                        bmp=(bm+abs(bm))/2.0;
                        bmm=(bm - abs(bm))/2.0;
                        % Leung & Qian (2006)
                        lhs=(app - amm)/DH+(bpp - bmm)/DH;
                        rhs=(amp*lam(k, h-1) - apm*lam(k, h+1))/
DH+…
                                (bmp*lam(k-1, h) - bpm*lam(k+1,
h))/DH;
                        % solve and update
                        if lhs>EPS_ADJ
                                lamt=rhs/lhs;
                        else
                                lamt=0.0;
                        end
                        lam(k,h)=min(lam(k,h), lamt);
                        % INF guard
                        if (TT(k,h)>=Inf) || (TT(k+1,h)>=Inf) ||
(TT(k-1,h)>=Inf) || …
                                (TT(k,h+1)>=Inf)   ||   (TT(k,h-
1)>=Inf)
                                lam(k,h)=0.0;
                        end
                end
                k=k+ndy;
        end
        h=h+ndx;
end
end
```

The 'sweep.m' is shown below:

```
function    [S,TT]=sweep(S,TT,nxsrc,    nysrc,    nx1,nx2,ndx,
ny1,ny2,ndy,NY,NX,DH)
% sweep over FD-grid
h=nx1;
```

```
for i=1:NX
        k=ny1;
        for j=1:NY
                % -- pick a,b from upwind neighbors (handle inte-
rior/borders/corners) --
                if   (h>1)   &&   (h<NX)   &&   (k>1)   &&   (k<NY)
% interior
                        a=min(TT(k, h-1), TT(k, h+1));
                        b=min(TT(k-1, h), TT(k+1, h));
                elseif    (h==1)     &&     (k>1)     &&     (k<NY)
% left border
                        a=min(TT(k, h), TT(k, h+1));
                        b=min(TT(k-1, h), TT(k+1, h));
                elseif    (h==NX)     &&     (k>1)     &&     (k<NY)
% right border
                        a=min(TT(k, h-1), TT(k, h));
                        b=min(TT(k-1, h), TT(k+1, h));
                elseif    (k==1)     &&     (h>1)     &&     (h<NX)
% top border
                        a=min(TT(k, h-1), TT(k, h+1));
                        b=min(TT(k, h), TT(k+1, h));
                elseif    (k==NY)     &&     (h>1)     &&     (h<NX)
% bottom border
                        a=min(TT(k, h-1), TT(k, h+1));
                        b=min(TT(k-1, h), TT(k, h));
                elseif (h==1) && (k==1)
upper-left corner
                        a=min(TT(k, h), TT(k, h+1));
                        b=min(TT(k, h), TT(k+1, h));
                elseif (h==1) && (k==NY)
lower-left corner
                        a=min(TT(k, h), TT(k, h+1));
                        b=min(TT(k-1, h), TT(k, h));
                elseif (h==NX) && (k==NY)
lower-right corner
                        a=min(TT(k, h-1), TT(k, h));
                        b=min(TT(k-1, h), TT(k, h));
                elseif (h==NX) && (k==1)
upper-right corner
                        a=min(TT(k, h-1), TT(k, h));
                        b=min(TT(k, h), TT(k+1, h));
                end
                % -- calculate solution --
                sh=S(k, h) * DH;
                if abs(a - b)>=sh
                        Tt=min(a, b)+sh;
                else
                        rad=2*(S(k,h)^2)*(DH^2) - (a - b)^2;
                        Tt=(a+b+sqrt(rad))/2;
                end
                % -- accept if smaller --
                TT(k, h)=min(TT(k, h), Tt);
                k=k+ndy;
        end
```

```
            h=h+ndx;
    end
    end

    --------------------------------------------------------------------------------
```

Reference

Zhao Y, Curtis A (2019) Relative source location using coda-wave interferometry: method, code package, and application to mining-induced earthquakes. Geophysics 84(3):F73–F84

The manufacturer's authorised representative in the EU is Springer Nature Customer Service Centre GmbH, Europaplatz 3, 69115 Heidelberg, Germany. If you have any concerns regarding our products, please contact ProductSafety@springernature.com

Printed and bound by CPI Group (UK) Ltd, Croydon, CR0 4YY
21/09/2026
02220015-0003